ENVIRONMENTAL MANAGEMENT AND GOVERNANCE

Recent attention has focused either on the global environment or on regulatory instruments for preventing environmental harm. This book addresses different aspects of environmental management that also raise fundamental questions about human actions and governmental roles.

There is a growing recognition of the failures of current environmental policy mandates and of the importance for environmental sustainability of decisions made by local governments about land use and development. At issue is how policies can be better designed and implemented to increase shared commitment between different layers of government to environmental goals and to foster practices that promote, rather than undermine, sustainability. Recognition of these challenges leads to consideration of a different form of policy entailing a shift from prescription and coercion to more flexible intergovernmental arrangements. Yet basic questions remain about the ability to bring about the desired intergovernmental cooperation and the efficacy of the approach: Does it foster stable partnerships between different levels of government? What choices are made about uses of land and development management? How do these choices affect sustainability?

The authors address these questions through detailed examination of "cooperative" and "coercive" intergovernmental regimes. They examine policy initiatives in New Zealand and Australia that empower local governments to devise methods for managing the environment in a sustainable way. These are contrasted with the more coercive and prescriptive approaches used in corresponding programs in the United States. The focus is on how the different regimes influence choices by local governments about land use and development in areas subject to natural hazards.

Covering important topics in land use, natural hazard management and sustainability, *Environmental Management and Governance* will prove invaluable to scholars and practitioners, including political scientists, policy makers, environmental managers, geographers, members of the broad community of hazard research and management, and all those interested in promoting sustainability.

D0377245

ENVIRONMENTAL MANAGEMENT AND GOVERNANCE

Intergovernmental approaches to hazards and sustainability

*Peter J. May, Raymond J. Burby,
Neil J. Ericksen, John W. Handmer,
Jennifer E. Dixon, Sarah Michaels
and D. Ingle Smith*

London and New York

First published 1996
by Routledge
11 New Fetter Lane, London EC4P 4EE

Simultaneously published in the USA and Canada
by Routledge
29 West 35th Street, New York, NY 10001

Routledge is an International Thomson Publishing company

Typeset in Garamond by Keystroke, Jacaranda Lodge, Wolverhampton
Printed and bound in Great Britain by TJ Press (Padstow) Ltd, Padstow, Cornwall

British Library Cataloguing in Publication Data
A catalogue record for this book is available from the British Library

Library of Congress Cataloguing in Publication Data
Environmental management and governance : intergovernmental approaches
to hazards and sustainability / Peter J. May . . . [et al.].
p.   cm.
Includes bibliographical references and index.
1. Land use—Government policy—Case studies. 2. Hazardous geographic
environments—Government policy—Case studies.
3. Central–local government relations—Case studies.
4. Environmental management—Case studies. I. May, Peter J.
HD156.E58 1996
363.7'056—dc20   95-52500

ISBN 0–415–14445–0 (hbk)
ISBN 0–415–14446–9 (pbk)

In memory of an unwavering companion named Yankee

P. M.

CONTENTS

ILLUSTRATIONS

PLATES

FIGURES

TABLES

NOTES ON AUTHORS

Peter J. May is Professor of Political Science at the University of Washington. He is the author of *Recovering from Catastrophes: Federal Disaster Relief Policy and Politics* (1985), co-author of *Disaster Policy Implementation: Management Strategies under Shared Governance* (1986), and has contributed articles to a variety of policy and planning journals. His recent research has addressed the design and implementation of regulatory policies for environmental problems.

Raymond J. Burby is DeBlois Chair of Urban and Public Affairs and Professor of Urban Planning at the University of New Orleans. Author or co-author of a variety of books on hazards management, including *Sharing Environmental Risks: How to Control Losses to Infrastructure from Natural Disasters* (1990), he has also contributed articles to a variety of planning journals on these topics. His research interests center on management of the impacts of urban development and on regulatory enforcement.

Neil J. Ericksen is Associate Professor and Director of the Centre for Environmental and Resource Studies at the University of Waikato, Hamilton, New Zealand, and has authored reports and articles for the Asia-Pacific region. His research focuses on environmental policy and planning, with particular reference to natural hazards and the human dimensions of climate variability and change.

John W. Handmer is Reader in Natural Hazards and Environmental Management at Middlesex University, London. Among other books, he co-authored *Risks and Opportunities: Conflict Resolution in Natural Resource Management* (1995) and has contributed articles to hazards and environmental management journals. His research centers on hazard and environmental management, with Australia and Britain as regional specialties.

Jennifer E. Dixon is Associate Professor in the Department of Resource and Environmental Planning, Massey University, Palmerston North, New Zealand, and her previous publications include articles in *Environmental Impact Assessment Review* and the *Journal of Environmental Planning and Management*. Her research focuses on environmental impact assessment and on planning practice.

Sarah Michaels is Assistant Professor in the Department of Urban and Environmental Policy at Tufts University, Massachusetts. She has contributed to *Policy Studies Review*, and her research interests center on comparative environmental policy, natural resources management, and the role of innovation in decision-making.

D. Ingle Smith is Senior Fellow in the Centre for Resource and Environmental Studies, Institute of Advanced Studies of the Australian National University. His research has been primarily in the field of water resources, specializing in the hazards of flood and drought with a regional focus on Australia.

PREFACE

Many books have been written about environmental problems and prospects for addressing them. Much of the recent attention has focused either on the global environment or on regulatory instruments for preventing environmental harms. The former deal with the big issues of sustainability and the health of the planet. The latter deal with issues of the quality of the environment. Each of these topics raises fundamental questions about the relationship of humans to the environment. They also raise issues about the role of governments in preventing environmental harms or averting catastrophes.

This book addresses different aspects of environmental management that also raise fundamental questions about human actions and governmental roles. We examine issues that arise when national or state governments try to persuade local governments to be good stewards of the environment. These are lesser studied problems in environmental management, but they are taking on new urgency. There is a growing recognition of the failures of current environmental policy mandates and of the importance for environmental sustainability of local government decisions about land use and development. Many argue that future directions for environmental policy include less emphasis on regulatory prescription and greater reliance on local governments as partners in pursuing paths to sustainable futures.

This requires a rethinking of the ways in which higher-level governments influence the decisions made by local governments. The United States, long looked to as the international leader in environmental policy, has failed to lead the way in rethinking the intergovernmental dimensions of environmental management. There is a growing consensus in the United States that federal and state environmental mandates are too heavy-handed. At issue is how policies can be better designed and implemented to increase shared commitment among different layers of government to environmental goals and to foster practices that promote, rather than undermine, sustainability. Recognition of these challenges leads to consideration of a different form of policy that we label as "cooperative intergovernmental policy." It entails a shift from policy prescription and coercion to more flexible intergovernmental arrangements.

Lacking clear-cut exemplars of such policies in the United States, we turn to other countries to draw lessons. Our comparisons are environmental management in New Zealand and hazard management in New South Wales, Australia. New Zealand has received world-wide attention for its reform of resource and environmental policy and its vision of integrated, "effects-based" environmental management. Although less comprehensive, the intergovernmental regime we study in New South Wales provides a noteworthy example of a shift from a heavy hand to a more flexible approach to help local governments cope with flood hazards. The lessons we draw are based on a comparison of the experiences with these policies and the experiences in Florida in the United States with its highly regarded, but highly coercive, approach to growth management.

The book is the product of a joint effort of an international group of researchers who brought differing perspectives to the endeavor. The project was directed by Peter J. May, a political scientist at the University of Washington. Investigators in each country took on responsibility for managing respective data collection. Raymond J. Burby, a planning scholar at the University of New Orleans, took responsibility for data collection in the United States. Neil J. Ericksen, a geographer at the University of Waikato, led a group of researchers in the data collection for New Zealand. His collaborators were Jennifer E. Dixon, a planning scholar in the Department of Resource and Environmental Planning at Massey University, and Sarah Michaels, a geographer at Tufts University. At the time, each was part of the Centre for Environmental and Resource Studies of the University of Waikato. John W. Handmer, a geographer at Middlesex University in England, collaborated with David Ingle Smith, a water resources specialist at the Australian National University, in collecting data about experiences in New South Wales, Australia. Both were associated with the Centre for Resource and Environmental Studies at the Australian National University.

We sought to write this book as an integrated account rather than as an edited volume. As a consequence, most of the chapters were written jointly by several members of the research group. Peter J. May conceptualized the overall book, served as lead author for four chapters, and collaborated in the authorship of several other chapters. Raymond J. Burby took the lead in authoring Chapter 2, which addresses Florida's growth management program, and Chapter 10 in addressing what we label the "commitment conundrum." He also collaborated in co-authoring Chapter 7, which examines local planning, compliance, and innovation. Neil J. Ericksen took the lead in authoring Chapter 3, which discusses experience in New Zealand, and Chapter 8 in considering sustainable management strategies. He also contributed to Chapter 6 on the role of regional governments in environmental management. John W. Handmer took the lead in authoring Chapter 4, which discusses experience in New South Wales, and contributed to the case-study material in Chapter 8. Jennifer E. Dixon collaborated in co-authoring Chapter 7, and contributed to the discussion of the New Zealand experience in Chapter 3 and the regional role in Chapter 6.

Sarah Michaels took the lead in co-authoring Chapter 6, which addresses the roles of regional government in environmental management. David Ingle Smith collaborated in co-authoring Chapter 9, which examines the outcomes of cooperative policies, and contributed to the case-study material in Chapter 8.

ACKNOWLEDGMENTS

This project involved the participation of many individuals without whom this book would never have been completed. Particularly important are those who provided especially able research or other assistance. Mark A. Donovan, a graduate student in political science at the University of Washington, assembled the combined datasets, helped with the analyses, and contributed to the appendix on methodological issues. Andrée Renée Jacques, a graduate student in the College of Urban and Public Affairs at the University of New Orleans, assisted in gathering information from Florida. Nick Sellers, a research assistant in the Centre for Resource and Environmental Studies at the Australian National University, supported the Australian investigators in various aspects of the surveys of local government and helped collect data for the profiles of development patterns for the selected communities. Claire Gibson, a research assistant in the Centre for Environmental and Resource Studies at the University of Waikato, assisted the New Zealand investigators in various aspects of the surveys of local governments and helped with gathering and analyzing the policies and plans of local government. Sarah Chapman, a graduate student in geography at the University of Waikato, helped with the analysis of the quality of plans and with case studies. Mark Dunlop, a graduate student in law, analyzed relevant statutes. Max Oulton, senior technician in the Department of Geography of the University of Waikato, produced the figures for the book.

The book was strengthened by the contributions of three scholars who joined us at the Rockefeller Foundation's Study and Conference Center in Bellagio, Italy, to discuss a draft of the book. These are Robert Nakamura, a political scientist at the State University of New York at Albany; Dennis Parker, a geographer and environmental scholar at Middlesex University in England; and Soeren Winter, a political scientist at Aarhus University in Denmark. We are especially grateful to them for their comments and to the Rockefeller Foundation for making it possible for our group to meet in a team residency.

A number of others were especially helpful in providing advice or commenting on key aspects of the book. Through her attention to the details of publishing this book, Sarah Lloyd of Routledge restored our faith in acquisitions editors. In his role as a program manager for United States National Science Foundation, William Anderson provided needed encouragement for the project at various stages. Philip Berke, a planning scholar at the University of North Carolina at Chapel Hill, provided advice about coding plans and designing the questionnaires used in New Zealand. Officials at various levels of government in all three settings gave generously of their time and insights in either answering detailed questionnaires, participating in face-to-face interviews, providing documents, or commenting on chapter drafts. We are especially grateful to these individuals, and in particular to: Neil Benning, Jim Bodycott, and Ian White in New South Wales; and Lindsay Gow, Dallas Bradley, Colin Gray, David Ray, and Warren Tuckey in New Zealand.

In addition to these individuals, we benefited from the comments of colleagues at research meetings and in the review of two articles that draw from parts of this book. A version of Chapter 5 appears as an article by Peter J. May, "Can Cooperation Be Mandated? Implementing Intergovernmental Environmental Management in New South Wales and New Zealand," *Publius: The Journal of Federalism* 25, no. 1 (Winter 1995): 89–113 under copyright by the Center for the Study of Federalism. A version of parts of Chapter 7 appears as an article by Peter J. May and Raymond J. Burby, "Coercive Versus Cooperative Policies: Comparing Intergovernmental Mandate Performance," *Journal of Policy Analysis and Management* 15, no. 2 (Spring 1996) under copyright by the Association for Public Policy Analysis and Management.

Much needed help with photos was provided by several individuals. Phil Flood of the Florida Department of Environmental Protection provided photos of hurricane damage in Florida, including the one used on the cover of the paperback. Yvette MacDonald made available photos from the collection of the Environmental Waikato Regional Council, New Zealand. The Australian photos were made available by P. D. May of the Disaster Awareness Program of Emergency Management Australia. Many of the New Zealand photos were provided by Neil Ericksen and Richard Warwick of the Centre for Environmental and Resource Studies, University of Waikato. We also thank Robert Deyle for his invaluable assistance in helping us obtain photos.

This project was made possible by financial support provided by a number of institutions. Primary funding was provided by the United States National Science Foundation under grant BCS-9208082 to the University of Washington. Some of the data for Florida are derived from research conducted by Raymond J. Burby and Peter J. May in collaboration with John DeGrove and his colleagues at the Center for Urban and Environmental Problems, a joint center of Florida Atlantic University and Florida International University. That research was supported by National Science Foundation grant BCS-8922346 to the University of North Carolina. Additional funding for travel and other expenses was provided by the Henry M. Jackson School of International Studies of the University of Washington and each of the participating universities. The Rockefeller Foundation provided support for our meeting at the Villa Serbelloni in Bellagio, Italy. The contents of this book are not necessarily endorsed by the National Science Foundation or other institutions that helped finance the research.

1

RETHINKING INTERGOVERNMENTAL ENVIRONMENTAL MANAGEMENT

The twenty-five years since the 1970 Earth Day have been sobering for environmental policy-making. By today's standards there was an almost naive faith in the power of government intervention as a means of limiting environmental harms. We now know that much more is involved in gaining compliance with environmental regulations than specifying a set of standards and associated penalties. We also know that the imperative in industrialized countries to promote economic progress has undermined longer-term sustainability. Recognition of the limitations of the environmental policies of the 1970s and 1980s has led to a questioning of the adequacy of existing environmental regulatory regimes.

The intergovernmental dimensions of environmental management for countries with multi-tiered systems of governance were also little understood twenty-five years ago. Direct national regulation of environmental harms bypassed thornier intergovernmental issues. In the intervening period, intergovernmental issues have taken on added significance. A newly restricted sense of the proper scope of national government in a number of industrialized countries, heralded by the election of reform-minded governments in the 1980s, set the stage for rethinking governmental roles. These roles were further challenged with concerns about governmental abuse of property rights under restrictive environmental regulations. At the same time, there has been increased recognition of the negative consequences of development and inappropriate uses of land for the quality of the environment and future environmental sustainability.[1] Addressing these consequences entails consideration of governmental regulation of development and land use.

While it was conceivable to devise policies addressing these aspects of environmental management so that they involved direct regulation by state or national governments, the political and practical realities for multi-tiered governmental systems lead to shared governance of these functions. In federalist systems of governance, state and local governments share these responsibilities. In unitary systems of governance, these functions are shared by national and local governments. The nature of the partnership is important. In the United States, a number of states have directed local governments to protect sensitive

1

environments or to plan for growth. In some instances, they have followed requirements imposed on the federal government. But, the state and federal requirements are perceived by many local governmental officials as being overly prescriptive and coercive.[2] They complain about the failure of higher-level governments to fund the costs of implementation, the lack of flexibility in the required actions, and the shifting to them of political blame for infringement of property rights. As a consequence of these concerns, local governments have been reluctant partners in the intergovernmental arrangements.

This reluctance creates a shared governance dilemma for the design of inter-governmental policies for environmental management.[3] The cooperation of local governments is essential for instituting appropriate development regulations and controls over land use. Yet, the requisite intergovernmental partnerships among state and local governments for these functions are potentially unstable and sometimes impossible to consummate. This is because state and local government officials do not always agree about the appropriateness of controls over development or land use for environmental purposes. State officials generally find themselves facing pressures to address environmental harms, and therefore want local governments to undertake these aspects of environmental management. In contrast, local officials generally find themselves facing strong pressures to promote development, rather than restrict it. They are often reluctant to impose the type of restriction that state officials desire. There is, of course, variation in these pressures that leads to differences among states in willingness to mandate growth management or other environmental controls.[4]

Key challenges for those thinking about these aspects of environmental management are to enhance the commitment of local government officials to addressing environmental problems and to strengthen their capacity to under-take needed actions. How this is best done is far from clear. The difficulties of mandating such action are underscored by the negative reactions of local officials in those states (or nations) that have enacted, or attempted to enact, strong requirements for local regulation of development and land use. Recognition of these difficulties leads to consideration of a more flexible form of inter-governmental policy mandates. Various authors have written about collaborative forms of environmental management under the labels "co-production," "collaborative planning," and "civic environmentalism."[5] Current ideological preferences for fewer dictates from above, recognition of the conflict among layers of government produced by restrictive mandates, and demands by local government officials for flexibility all point to future emphasis on inter-governmental collaboration.

RETHINKING INTERGOVERNMENTAL ENVIRONMENTAL MANAGEMENT

This book examines what we label "cooperative intergovernmental policy" as an approach to fostering stronger environmental management by local governments and as a means for enhancing sustainability. Although this policy approach appears to have much promise in overcoming the limitations of the prescriptive and coercive approach, basic questions remain about the ability to bring about the desired cooperation and the efficacy of the approach: Does it foster stable partnerships among different levels of government? Does it enhance efforts by local government to manage the environment? What choices are made about uses of land and development management under this intergovernmental regime? How do these choices affect sustainability? The central tasks of the book are to address these questions. This involves consideration of exemplars of this approach and the lessons that can be drawn from these experiences. The conceptual foundations for the book are provided in the remainder of this section.

Coercive and cooperative intergovernmental policies

A distinction between coercive and cooperative forms of intergovernmental policy mandates is central to this book.[6] Each mandate entails the imposition of procedural and/or substantive requirements by a state (or national) government on local governments, either as conditions for assistance or as direct orders. Each approach can in theory be used to enhance the adherence of local governments to higher-level policy objectives. But, the designs differ in terms of their underlying assumptions and use of policy tools. Table 1.1 depicts the key distinctions.

Coercive intergovernmental mandates treat local governments as regulatory agents charged with following rules prescribed by higher-level governments. These mandates spell out detailed standards and procedures for achieving policy goals, thereby reducing state or local discretion in policy development. Sanctions are applied when governments fail to undertake their prescribed roles or deviate from procedural prescriptions of mandates. Coercive mandates pay some attention to building the capacity of local governments to comply, but that is secondary to putting in place monitoring systems for compliance and invoking penalties for noncompliance.

Cooperative intergovernmental mandates try to enhance local government interest in and ability to work toward achieving higher-level policy goals. Local governments act as regulatory trustees in seeking appropriate means to meet goals they share with higher-level governments. These regimes may prescribe planning or process elements to be followed (a form of policy mandate), but they do not prescribe the particular means for achieving desired outcomes. Cooperative mandates use financial and technical assistance for the dual purpose of enhancing the commitment of local governments to policy goals and increasing their capacity to act.

The compelling logic of each form of intergovernmental mandate is very different. Coercive policies are highly paternalistic. The logic is that mandating governments know the appropriate actions to be taken by local governments. Given the reluctance of the latter, the approach is to compel the desired actions. Cooperative intergovernmental policies are less paternalistic. The logic is that

3

Table 1.1 Intergovernmental policy designs

Features	Comparison of policy features	
	Coercive policy design	*Cooperative policy design*
Role of lower-level governments (state, regional, or local)	*Regulatory agents:* Enforce rules or regulations prescribed by higher-level governments.	*Regulatory trustees:* Develop and apply rules that are consistent with higher-level goals.
Emphasis of intergovernmental mandate	Prescribe regulatory actions and process. Specify regulatory actions and conditions, along with required process or plans.	Prescribe process and goals. Specify planning components and considerations, along with performance goals.
Control of lower-level governments	Monitoring for procedural compliance. Enforcement and sanctions for failing to meet deadlines, for not adhering to prescribed process, or for not enforcing prescribed rules.	Monitoring for substantive compliance with more limited monitoring for procedural compliance. Monitoring systems for assessing outcomes and progress toward them.
Assumptions about intergovernmental implementation	Compliance is a potential problem. Need for uniformity in application of policies.	Compliance is not a problem. Need for local discretion in policy development.
Source of policy innovation	Higher-level governments	Lower-level governments
Implementation emphasis	Inducing adherence to policy prescriptions and regulatory standards. Building "calculated" commitment as a primary means of inducing compliance.	Building capacity of subordinates to reach policy goals. Enhancing "normative" commitment as a primary means of inducing compliance.

higher-level governments know that action must be taken, but there is uncertainty as to what that best constitutes. Local governments are told to think seriously about the problems and their solutions following prescribed planning processes, but the specific actions are left to the local governments to determine.

The two forms of intergovernmental policies differ in their assumptions about local government commitment to policy goals and about capacity to carry out actions in order to reach those goals. The coercive design recognizes the fundamental tensions of shared governance, and thus it presumes there will be conflicts among layers of government over goals or means for reaching goals. The coercive solution is to apply sanctions to recalcitrant governments. Compliance is achieved partly as a result of calculations that local governments make about the consequences of failing to comply; a set of decisions that we label calculated commitment.

4

The cooperative design assumes local governments do not have any fundamental disagreements with policy aims and therefore do not have to be forced to comply. They are assumed to already possess at least a modicum of commitment to policy goals, which we label as normative commitment. This consists of a shared willingness to work toward objectives, as opposed to the fear of reprisal that drives calculated commitment.[7] But normative commitment needs to be mobilized, for which financial and other inducements can be important. By removing barriers created by deficiencies in capacity and by enhancing normative commitment, cooperative intergovernmental policies seek local government compliance with the objectives of higher-level governments.

Reflecting the different logic of the two mandates, they also differ in the way they foster new approaches to planning, new ideas about regulatory instruments, or the re-combination of regulatory provisions to better match local circumstances. We consider these to be different aspects of innovation in environmental management. Coercive intergovernmental policies impose innovation from above. The burden is on higher-level governments to devise better ways, which then can be implemented by local governments. Cooperative intergovernmental policies seek innovation from below in allowing opportunities for experimentation by local governments. The burden is on local governments to figure out appropriate solutions and to make adjustments if they prove problematic.

There are many conceivable variants of these two polar extremes of coercive and cooperative policy designs. For each policy type, there can be variation in the extent to which the process is prescribed and in the extent to which the related actions or outcomes are prescribed. We label the former procedural prescription and the latter substantive prescription. Mandating governments can monitor for compliance with either or both forms of prescription. Coercive policies vary in the stringency of sanctions in seeking calculated compliance and in the use of inducements to head off potential backlash. Cooperative policies vary in the extent of inducements to collaborate.

Intergovernmental policies and sustainability

The critical test for intergovernmental environmental management policies is whether they foster wise choices by governments about development and growth. These choices can profoundly affect the economic base of communities, the quality of the environment, and the vulnerability of communities to natural hazards. The stakes in making these decisions are often large, and conflicts over the outcomes are not easily resolved. Because state (or national) policy mandates affect – or even prescribe – these choices, the design and implementation of intergovernmental mandates are important.

A key issue is what constitutes criteria for evaluating wise choices about land use and development. The Bruntland Report of 1987 called worldwide attention to sustainable development as an objective for environmental policy. The report defined sustainable development as actions taken in order to meet present needs without compromising the ability of future generations to meet their needs.[8] Sustainable development is a broad concept that goes beyond the tradition of wise use of resources and introduces ecological, economic, equity, and social considerations. Some argue that this is an unrealistic objective and that other conceptualizations of sustainability provide more appropriate objectives. For example, New Zealand policy-makers have established a goal of the sustainable management of natural and physical resources. Under this concept, the emphasis is on managing resources so that the resource needs of future generations can be met. A different concept is that of environmental sustainability, which places emphasis on protecting the environment and suggests restrictions on actions that are harmful to the environment. In this book, we use the general term sustainability to refer to the set of inter-generational objectives with respect to the environment that are suggested by these concepts (while recognizing that the concepts are not interchangeable). We employ more specific terms – sustainable development, environmental management, or environmental sustainability – when referring to different aspects of sustainability that have been incorporated into the policies we consider.

Fostering local government consideration of sustainability provides fundamental challenges for intergovernmental policy design. The substantive challenges relate to the complexities of human interactions with the natural environment. Policies are sought that mediate ecological fluctuations and

Plate 1.1 Unsustainable Florida coastal development in the aftermath of Hurricane Opal, 1995 (courtesy of Phil Flood, Florida Department of Environmental Protection)

perturbations in the natural environment for which a central difficulty is the ability to diagnose and assess the implications of such change. Appropriate policy actions that are based on integrated information about environmental effects and other outcomes are sought. This entails consideration of the cumulative effects of development among different environmental media (e.g., air, land, water) and among neighboring jurisdictions. The latter raises issues about the appropriate scale of policy-making for which a case can be made for a regional government role in providing the desired integration within defined natural boundaries. However, such regional approaches to environmental management have proved to be problematic when attempted on a widespread basis.

Other difficulties in fostering attention to sustainability are also relevant to this book. One difficulty is devising policies that will be able to survive major political and ideological challenges. Having sustainable policies is itself a key to the type of long-term vision that environmental sustainability requires. Another difficulty is fashioning policies that are flexible and goal oriented in recognition that there are many different pathways to the goal of sustainability. Taken together, these considerations argue against the type of rigid policy design and detailed policy prescription found in coercive intergovernmental policies. They argue for flexibility, the ability to accommodate change, and a willingness to negotiate – each of which are elements of cooperative intergovernmental policies. The potential of this policy approach for enhancing local government choices about environmental management and sustainability is why the cooperative policy approach is a central focus of this book.

CONSIDERING CROSS-NATIONAL COMPARISONS

Assessing the implications of different forms of intergovernmental mandates for environmental management requires development of an understanding of how each plays out in practice. From our prior research about state mandates in the United States, we are able to identify more or less consistent examples of coercive intergovernmental policy designs, but we fail to find clear-cut examples of cooperative policy approaches.[9] We turn to other countries for two exemplars of this approach to intergovernmental environmental management. These are environmental management policy in New Zealand and floodplain management policy in the state of New South Wales, Australia. The coercive intergovernmental approach is illustrated by what many observers consider to be an exemplar – Florida's growth management program.

The selection of these experiences as examples of different approaches to environmental management is based on several considerations. One is the desire to draw a contrast between the two forms of policy mandates as they apply to environmental management. The lack of sufficient contrast in the approaches employed in the United States led to consideration of experiences in other countries. A second consideration is the desire to have comparable settings so that extraneous factors do not confound inferences about different policy types. We address the comparability of settings later in this chapter. A third consideration is a practical one of choosing settings where we could assemble a research team. This is particularly important for the detailed data collection that we sought. The remainder of this section provides an overview of the policy exemplars.

Coercion and policy prescription in Florida

The experience in Florida serves as a strong example of the type of coercive and prescriptive intergovernmental policy found in the United States. It provides a key comparison for experiences with the cooperative approaches to environmental management in this book. The Florida experience has been widely cited in the American planning literature as "the example" of an effective state policy for stimulating efforts by local government to manage growth and reduce environmental harms. The trend in revising Florida's policies has been one of adding greater coercion with respect to local governments, providing a sharp contrast to the trends in New Zealand and in New South Wales.

Beginning in the mid-1960s, policy-makers in Florida enacted a series of state policies governing development in coastal and other hazard-prone areas. A strong state role was established from the outset. Provisions of coastal legislation required state permits for specified development. Environmental planning legislation required state review of development proposals within "areas of critical state concern" or for developments with regional impacts. By the mid-1970s, broader growth and planning concerns led to the enactment of a new set of comprehensive state and regional planning requirements that specified detailed requirements for state, regional, and local planning. A major reform of the comprehensive planning legislation in 1985 added requirements for greater consistency in planning activities (both among levels of government and across local jurisdictions) and increased sanctions for the failure of local governments to adhere to the planning requirements and deadlines. The additional regulatory teeth of the 1985 legislation evidenced a strong move toward greater intergovernmental coercion.

The approach to environmental and comprehensive planning in Florida is unique among American states in the toughness in approach to local governments. As a consequence, it is a good case for examining the influence of coercive intergovernmental policy designs. The environmental legislation of most states has strong provisions prescribing conditions for private development in environmentally sensitive areas, but contain weaker provisions concerning actions of local governments in carrying out state mandates. Florida's legislation differs in that there are strong sanctions for local governments that fail to adhere to required planning provisions. There is a serious state review of local plans and development regulations for consistency with state guidelines, and those jurisdictions with deficient plans and regulations are required to make changes.

The evolution of environmental and growth management legislation in

Florida embodies a wrestling over time with different notions of sustainability. The restrictions on development in critical areas and along the coast that were put in place in the 1960s and 1970s entail a strong emphasis on environmental sustainability. Responding to a variety of social, economic, and growth pressures, the more recent growth management policy increases the role of those considerations in local planning thereby drawing attention to sustainable development considerations.

Reform and cooperative policy in New Zealand

The cooperative environmental management approach in New Zealand is embodied in an extensive set of reforms in local governmental structure and resource management legislation undertaken from 1987 through 1991. These reforms were a conscious effort to restructure local governments and to devolve power from central government to the local level. The *Resource Management Act* adopted in 1991 replaced nearly sixty pieces of environmental and resource management legislation with a single, comprehensive approach to natural resource and environmental management. In choosing this policy for study, we concur with the observations that: "A set of reforms more significant for environmental policy beyond New Zealand – whether in terms of theoretically based insight into strategic design or practical experimentation with institutional arrangements – can scarcely be imagined."[10]

The *Resource Management Act* establishes a national framework and goals for resource and environmental management and creates a partnership among regional and local governments for environmental planning and management. Under this partnership, the central government establishes national goals and policy statements; regional governments develop regional policy statements and plans; and local governments prepare district land-use plans and development rules. The regulatory philosophy is one of greater flexibility and reticence to impose minimum standards. The aim is to allow regional and local governments to set rules that are appropriate to local circumstances, as long as they do not contradict national goals. The policy guidelines provided to local and regional governments state that they are "free to develop their own approaches as long as they achieve the outcomes specified in the Act and follow the specified process."[11] Efforts to increase the capabilities of local governments and their normative commitment to the policy consist of grants (originally only at the regional level), public education, and review and commentary about the policies or plans prepared by regional and local governments.

A noteworthy aspect is the emphasis that is placed on the overarching goal of sustainable management. This is defined by the legislation as "managing the use, development, and protection of natural and physical resources" to meet the reasonably foreseeable needs of future generations, safeguard the life-support capacity of various natural systems (e.g., air, water), and to avoid or mitigate adverse effects of human activities on the environment. This emphasis on environmental protection is different from the concept of sustainable development and its inclusion of social and economic development goals. As explained by New Zealand's Minister for the Environment: "[T]he Act is *not* about directing the wise use and development of resources in order to effectively promote and safeguard health, safety, convenience, and economic, cultural, and social welfare (to use the language of the [prior] Town and Country Planning Act). . . . Nor is it about the balancing between socio-economic aspirations and environmental concerns . . . [It] is first and foremost an environmental statute."[12]

Toward cooperative policy in New South Wales, Australia

Steps toward cooperative policy in New South Wales consist of a state flood policy, adopted in December 1984, that replaced controversial features of a more coercive intergovernmental policy that existed from 1977 to 1984. The change from a coercive to a more cooperative intergovernmental policy makes this a particularly useful case for study. The experience in New South Wales also provides an especially appropriate comparison with the experience in Florida, given that the latter entailed a move in the opposite policy direction over a similar time period.

The flood policy in New South Wales sets forth a "merits" approach to be followed by local governments when dealing with development in flood-prone areas.[13] Under this policy, local governments are required, as conditions for subsequent funding for flood mitigation and as a basis for establishing governmental immunity from legal liability for flood losses, to develop floodplain management plans and rules about development that lessen flood loss. The "merits" descriptor derives from the absence of state prescription about local decisions about proposed development, which the policy states are to be considered by local governments on their merits within the context of locally developed rules. The collaborative intent of the flood policy is evident from the language officials use to describe it: "The [Flood] Policy conforms closely with the [Environmental Planning and Assessment] Act in promoting the sharing of responsibility between the different levels of government with local planning being the responsibility of Local Government."[14]

Several features of cooperative policy designs are evident in this approach. The policy requires completion of a planning process, rather than particular actions for managing floods as conditions for future aid and waivers of immunity. The policy emphasizes goals that are presumably shared by the state and local governments, rather than prescribed standards to be met. Local governments can devise the best means within their communities for reaching the goals. A key feature of the policy design is building the capacity of local governments to undertake floodplain management.

The policy incorporates key aspects of sustainable development by requiring that decisions about flood policy and development be based on the merits of development in floodplains while taking into account social, economic, and ecological considerations. That is, local governments are asked to explicitly consider a range of issues involved in managing development. The intent is not to preclude development. The policy explicitly states that "flood-prone land is often a valuable resource for which preclusion of all forms of development is not necessarily the best policy."

Although the flood policy clearly signals new directions for local environmental management and intergovernmental cooperation, the broader aspects of environmental policy in New South Wales are governed by a more prescriptive and potentially coercive system of planning and land-use regulation. The *Environmental Planning and Assessment Act* of 1979 sets forth a hierarchical planning system that prescribes the content of local development management rules and reserves for the state the power to intervene in local development planning and approvals. Because these provisions govern state and local planning more generally, the flood policy is best depicted as an overlay of cooperative features to a more traditional, prescriptive policy entailing strong central government authority.

REFINING THE CROSS-NATIONAL COMPARISONS

Cross-national comparisons of policies introduce a host of potential problems in that differences in scale, legal structures, political cultures, administrative arrangements, and governance can confound the comparisons.[15] In choosing selected experiences at the state level in Australia and the national level in New Zealand, we think that the similarities with our state-level comparison in the United States far outweigh the differences. Nonetheless, there are fundamental considerations about the comparisons that shaped the research for this book.

Natural hazards as a topical focus

Each of the policy exemplars from Florida, New Zealand, and New South Wales addresses aspects of environmental management. However, they differ in scope and foci. Florida's growth management policy addresses the environmental and social consequences of growth. New Zealand's environment and resource management reforms address a range of physical and natural resources. In contrast, the New South Wales policy has a relatively narrow focus on flood management. Given the differences in scope and foci of these policies, we found it necessary to identify common elements for study. To provide this, we choose to focus on the natural hazards components of each policy mandate. This provides a focus for the comparison and a tractable set of environmental management issues to study.

The management of natural hazards embodies the more general aspects of environmental management addressed earlier in this chapter. Like many other environmental harms, human activity is a key source of problems. The effects of earthquakes, floods, and other hazardous events are as much a function of human activities as they are of the natural events.[16] By locating development in hazard-prone areas, building structures that fail in natural disasters, and depleting the resilience of natural processes, humans both contribute to increased risk and modify natural hazards in undesirable ways. Efforts by higher-level governments to influence these choices are important components of this aspect of environmental management, since few local governments would undertake extensive efforts to alter these activities without some form of external inducement or prodding.

The risks posed by natural hazards can be viewed as part of a broader class of environmental harms that constitute "public risks." These constitute such risks as ozone depletion and sea-level rise. Peter Huber, an authority on regulation of risks, characterizes these as risks that are "centrally produced or mass-produced, broadly distributed, often temporally remote, and largely outside the individual risk bearer's direct understanding and control."[17] These present collective action problems because citizens and public officials perceive them as having low probabilities of occurrence. As a consequence, there are limited incentives for private action and it is difficult to sustain broad public responses. Governmental intervention is required to address these risks, which in turn raises issues posed in this book concerning the degree of paternalism and the appropriate form of intervention.

Local government choices about the management of natural hazards raise

Plate 1.2 Housing at risk along steep hillsides in Wellington, New Zealand (photo by Neil Ericksen)

issues about economic development and environmental protection that are at the heart of broader debates about environmental management and sustainability. Decisions made by local governments about appropriate uses of land, controls over development, or regulation of construction in hazardous areas can have profound effects on economic vitality, environmental quality, and vulnerability to natural events. Like other aspects of environmental management, the stakes in making these decisions are often large, and conflicts over the outcomes are not easily resolved. There are few incentives for local officials to address these issues. There are not strong public constituencies pressing for policies to mitigate natural hazards.[18] Moreover, the costs of policy interventions tend to be up front and concentrated, whereas the benefits are often delayed and diffuse. These factors contribute to the shared governance dilemma that is at the heart of intergovernmental environmental management.

Efforts by higher-level governments to influence local government choices about the management of development in hazard-prone areas embody key challenges posed more generally by intergovernmental environmental mandates. One challenge is finding ways of dealing with problems that arise when local governments do not share higher-level policy objectives. Other challenges entail dealing with the complexities of implementation, which are compounded when multiple layers of government are involved.

Comparability of settings

The settings of the policy exemplars of this book have a number of similarities. In all three countries, local governments have primary roles in land-use control and development management. Each of the countries has a strong tradition, based on English law, of private property rights combined with governmental authority for protection of public welfare and safety. Interest groups are important political forces in shaping national and state legislation in each country, with the multi-party electoral structure in Australia and New Zealand providing a direct electoral link to environmental groups. Each of the countries elected reform-minded governments in the 1980s that helped to set the stage for considering new approaches to environmental management.

The differences in scale with respect to numbers of affected jurisdictions and populations would be immense if we attempted to compare national-level policies in the United States with the other settings. Our decision to compare experiences with aspects of Florida's growth management program in the United States with corresponding policy for the state of New South Wales, Australia, and national policy in New Zealand makes the scale more comparable. New South Wales has about the same land area as Texas, but with only a third of the population. New Zealand has roughly the population and land area of Oregon. Florida is the largest of these settings in population, but has a land area that is about one-half that of New Zealand and one-fifth that of New South Wales.

The choice of study sites also reflects efforts to control for differences in governmental structures. Australian federalism has a tradition of strong state-level governance among the six states and two territories with, until recently, relatively limited Commonwealth intervention in state affairs. Like the United States, Australian states have responsibility for regulation of land use and development. But, they too vary considerably in the scope and style of environmental and land-use policy.[19] In both countries, state-level environmental policies entail a mix of direct state regulation (e.g., state permits for activities in coastal zones) and reliance upon local governments as regulatory intermediaries. In recent years, state policy-makers in both Florida and New South Wales have had to grapple with the challenges posed by growth in urban areas, threats to environmental and coastal resources, and risks posed by natural hazardous events. The trend in both settings is one of stronger state controls with respect to environmental issues leading to a mix of state direct intervention and efforts to shape local planning processes and decisions.

New Zealand presents a different situation because it is not a federalist system comprised of a separate layer of state governments.[20] Until the late-1980s, New Zealand government authority was highly centralized with local governments having important, but subsidiary infrastructure and service roles. Various regional authorities existed for special purposes. A series of governmental reforms culminated in 1989 with amendments to the *Local Government Act*. These reforms, with subsequent adjustments, created regional authorities and collapsed local governmental authority into seventy-three local governments. The adjustments included reconstituting a few regions and local governments into unitary authorities with combined functions. These governmental reforms set the stage for a sharing of environmental management responsibilities among the newly created regional governments and the reconstituted local governments. This means for comparative purposes the New Zealand governance structure looks much like a state in the United States, but with a stronger regional government role in environmental management than found in any state. Unlike the situation in Florida and New South Wales, the trend has been a distinct one of devolution of decision-making away from central government to regional and local levels.

Methodological notes

One of our goals in thinking about the research design was to move beyond the descriptive comparisons of most cross-national studies to provide a more refined set of analytic and empirical comparisons. The statistical and case-study material within the book provide a sound basis for thinking about new directions for intergovernmental environmental management. The details of data collection and analytic methods are provided in an appendix.

Our understanding and analysis of the policy experiments reported in this book are based on several sources. These include information about the policies within each setting; state and central government agency efforts to implement the policies; local government choices about the management of natural hazards and the efforts they put into them; and for selected settings, changes in risk profiles. We also collected data comparing local government practices before and after introduction of the cooperative and coercive regimes, and data comparing the perceptions of local government officials about the character of intergovernmental relations under each regime.

This comparative information provides a basis for commenting about the consequences of cooperative intergovernmental regimes. We do this through statistical analyses of our survey data and qualitative observations from our case studies of experiences of local governments under these regimes. The former provide a basis for comparing experiences while taking into account key differences in settings. The latter provide an important basis for interpreting the statistical findings and for extending our understanding of policy impacts. The main caveat about the data is that at the time of our fieldwork in late 1993, New Zealand was only in the early stages of implementing the environmental management and associated reforms that were initiated in the early 1990s. As such, our comments about experience in New Zealand are necessarily preliminary.

A LOOK AHEAD

Part I of the book, comprising the next three chapters, provides a detailed examination of the evolution of the experiments in aspects of environmental management policy in the United States, New Zealand, and Australia that are the foci of this book. These chapters discuss the struggles that policy-makers faced in establishing and refining these intergovernmental regimes. The policies evolved in different ways in response to shortfalls in governmental efforts to manage development and its consequences.

Part II of the book contains five chapters in which we examine the promise of cooperative intergovernmental policies. Where relevant, comparisons are made with the coercive approach. Assessing the promise entails considering questions about policy implementation and outcomes: What is involved in bringing about intergovernmental partnerships? Can regional governments play a noteworthy role, given their precarious standing as governmental entities? Do local governments respond differently in preparing plans under cooperative and coercive regimes? Do cooperative policies enhance local government efforts to manage the environment? And, what are the outcomes of cooperative environmental policies? These are the central questions addressed respectively in the chapters in this part of the book.

Part III of the book provides observations about cooperative intergovernmental environmental regimes. This includes discussion of the longer-term potential for cooperative intergovernmental policies, and of the general applicability of key features of such policies. Recognition of the limitation of these policies leads to consideration of potential improvements in policy design and implementation.

NOTES

1 For overviews of these trends see: National Commission on the Environment, *Choosing a Sustainable Future* (Washington, D.C.: Island Press, 1993), pp. 113–118; and Ian Burton, Robert W. Kates, and Gilbert F. White, *The Environment as Hazard*, 2nd edition (New York: Guilford Press, 1993), pp. 219–240.

2 Relevant depictions of these relationships include: U.S. Advisory Commission on Intergovernmental Relations, *Federal Regulation of State and Local Governments: The Mixed Record of the 1980s* (Washington, D.C.: Advisory Commission on Intergovernmental Relations, 1993); U.S. Congress, General Accounting Office, *Federal–State–Local Relations, Trends of the Past Decade and Emerging Issues*, Report HRD-90-34 (Washington, D.C.: General Accounting Office, 1990).

3 For a discussion of this dilemma in the context of policies for natural hazards see: Peter J. May and Walter Williams, *Disaster Policy Implementation, Managing Programs Under Shared Governance* (New York: Plenum Press, 1986). More generally see: Daniel Elazar, American Federalism, A View from the States, 3rd edition (New York: Harper and Row, 1984); and Robert P. Stoker, "A Regime Framework for Implementation Analysis: Cooperation and Reconciliation of Federalist Imperatives," *Policy Studies Review* 9, no. 1 (Autumn 1989): 29–49.

4 For discussion of this variation among American states see: Scott A. Bollens, "State Growth Management: Intergovernmental Frameworks and Policy Objectives," *Journal of the American Planning Association* 58, no. 4 (Autumn 1992): 454–466; John DeGrove, *Planning and Growth Management in the States* (Cambridge, MA: Lincoln Institute of Land Policy, 1992); Jon A. Kusler, *Regulating Sensitive Lands* (Cambridge, MA: Ballinger Publishing, 1980); and Peter J. May, "Analyzing Mandate Design: State Mandates Governing Hazard-Prone Areas," *Publius, The Journal of Federalism* 24, no. 2 (Spring 1994): 1–16.

5 See: Scott A. Bollens, "Restructuring Land Use Governance," *Journal of Planning Literature* 7, no. 3 (1993): 211–226; David R. Godschalk "Implementing Coastal Zone Management 1972–1990," *Coastal Management* 20, no. 3 (1992): 93–116; John DeWitt, *Civic Environmentalism: Alternatives to Regulation in States and Communities* (Washington D.C.: Congressional Quarterly Press, 1994); and David M. Welborn, "Conjoint Federalism and Environmental Regulation in the United States," *Publius: The Journal of Federalism* 18, no. 1 (1988): 27–43.

6 This discussion draws from Peter J. May and John W. Handmer, "Regulatory Policy Design: Cooperative versus Deterrent Mandates," *Australian Journal of Public Administration* 51, no. 1 (March 1992): 43–53. The regulatory design and enforcement literature addressing regulation of the private sector makes related distinctions from which we generalize to the intergovernmental setting. For an overview see: Robert A. Kagan, "Regulatory Enforcement," in *Handbook of Regulation and Administrative Law*, ed. David H. Rosenbloom and Richard D. Schwartz (New York: Marcel Dekker, 1994), pp. 383–422. For discussion of intergovernmental mandates that relate to the definition used here see: Catherine Lovell and Charles Tobin, "The Mandate Issue," *Public Administration Review* 41, no. 3 (May/June 1981): 318–331; and U.S. Advisory Commission on Intergovernmental Relations, *Mandates: Cases in State–Local Relations* (Washington, D.C.: Advisory Commission on Intergovernmental Relations, 1990).

7 For related distinctions see: Raymond J. Burby and Robert G. Paterson, "Improving Compliance with State Environmental Regulations," *Journal of Policy Analysis and Management* 12, no. 4 (Fall 1993): 753–772; and Margaret Levi, *Of Rule and Revenue* (Berkeley and Los Angeles: University of California Press, 1988), pp. 48–70.

8 The Bruntland Report, named after the chair of the commission that prepared it, is published as: World Commission on Environment and Development, *Our Common*

Future (Oxford: Oxford University Press, 1987). Relevant discussion of the concepts of sustainability include: John A. Dixon and Louise A. Fallon, "The Concept of Sustainability: Origins, Extensions, and Usefulness for Policy," *Society and Natural Resources* 2, no. 2 (1989): 73–84; Robert Goodland, "Environmental Sustainability and the Power Sector," *Impact Assessment* 12, no. 3 (Fall 1994): 275–304; Timothy O'Riordan, "The Politics of Sustainability" in *Sustainable Environmental Management, Principles and Practice*, ed. R. Kerry Turner (London: Belhaven Press, 1988): pp. 30–50; and Michael Redclift, "The Multiple Dimensions of Sustainable Development," *Geography* 76, no. 1 (1991): 36–42.

9 The American state experience is reported in Raymond J. Burby and Peter J. May with Philip R. Berke, Linda C. Dalton, Steven P. French, and Edward J. Kaiser, *Making Governments Plan: State Experiments in Managing Land Use* (Baltimore: Johns Hopkins University Press, 1996).

10 Ton Bührs and Robert V. Bartlett, *Environmental Policy in New Zealand: The Politics of Clean and Green?* (Auckland: Oxford University Press, 1993), p. 113.

11 New Zealand Ministry for the Environment, "Regional Policy Statements and Plans, Guidelines" (Wellington: Ministry for the Environment, September 1991), p. 5.

12 Keynote speech of Honourable Simon Upton, Minister for the Environment, to the Resource Management Law Association, 7 October 1994 (Wellington: Ministry for the Environment, 1994), p. 3. Emphasis in the original.

13 The policy is referred to by different names among state government documents – "Floodplain Management Policy," "Flood Policy," and "Merits Flood Policy."

14 New South Wales Government, *Floodplain Development Manual*, Document PWD 86010 (Sydney: Public Works Department, New South Wales State Government, 1986), p. 3.

15 See: Ariane B. Antal, Meinolf Dierkes, and Hans N. Weiler, "Cross-national Policy Research: Traditions, Achievements, and Challenges," in *Comparative Policy Research, Learning from Experience*, eds Meinolf Dierkes, Hans N. Weiler, and Aarian Berthoin Antal (Berlin: WZB-Publications, 1987), pp. 13–49; Jurgen Feick, "Comparing Comparative Policy Studies – A Path Towards Integration?" *Journal of Public Policy* 12 (Part 3, July–September 1992): 257–285; and Sheldon Kamieniecki and Eliz Sanasarian, "Conducting Comparative Research on Environmental Policy," *Natural Resources Journal* 30, no. 2 (Spring 1990): 321–339.

16 Research in this tradition is summarized in: Ian Burton, Robert W. Kates, and Gilbert F. White, *The Environment as Hazard*, op. cit.

17 Peter W. Huber, "The Bhopalization of American Tort Law," in *Hazards: Technology and Fairness*, National Academy of Engineering (Washington, D.C.: National Academy Press, 1986), pp. 89–110, at 90. Also see: Peter J. May, "Addressing Public Risks: Federal Earthquake Policy Design," *Journal of Policy Analysis and Management* 10, no. 2 (Spring 1991): 263–285.

18 The classic treatment of a lack of a constituency for hazard mitigation is provided by Peter H. Rossi, James D. Wright, and Eleanor Weber-Burdin, *Natural Hazards and Public Choice: The State and Local Politics of Hazard Mitigation* (New York: Academic Press, 1982).

19 Commentary about Australian federalism and environmental management includes: Bruce Davis, "Environmental Management" in *Intergovernmental Relations and Public Policy*, ed. Brian Galligan, Owen Hughes, and Cliff Walsh (Sydney: Allen and Unwin, 1991), pp. 146–162; and Elim Papadakis, "Environmental Policy," in *Hawke and Australian Public Policy*, ed. Christine Jennett and Randal G. Stewart (South Melbourne, Macmillan of Australia, 1990), pp. 339–355.

20 The political context for New Zealand environmental policy is discussed in Ton Bührs and Robert V. Bartlett, *Environmental Policy in New Zealand: The Politics of Green and Clean?*, op. cit.

Part I

POLICY INNOVATIONS IN THE UNITED STATES, NEW ZEALAND, AND AUSTRALIA

INTRODUCTION

The environmental and hazard-management programs discussed in this book are exemplars of approaches that state or national governments have employed in efforts to persuade local governments to be good stewards of the environment. Florida's state mandate that local governments develop growth and environmental management plans serves as an illustration of a coercive and prescriptive approach to influencing decisions by local governments about development and land use. New Zealand's rewriting of environmental legislation and accompanying governmental reforms are dramatic attempts to bring about a collaborative partnership among regional and local authorities for resource and environmental management. These reforms have attracted international attention. The evolution of flood policy in New South Wales, Australia, provides an understanding of cooperative approaches to flood management embedded within a more traditional approach to environmental planning.

Part I of the book discusses the struggles that policy-makers faced in establishing and refining these intergovernmental regimes. The chapters depict the evolution within each setting of innovations in approaches to environmental management and in thinking about ways to promote sustainability. The policies evolved in response to shortfalls in previous efforts to manage development and its consequences. The specifics of how this played out varied among the settings with some choosing to pursue comprehensive approaches and others more focused approaches. The understanding of the policy evolution provided in Part I provides the foundation for examining experiences under each type of environmental management regime that is undertaken in subsequent parts of the book.

2

COERCION AND
PRESCRIPTION
Growth management in Florida

Several decades have passed since state governments in the United States began experimenting with programs to solve environmental problems by managing development. A plethora of single-purpose development regulations are on the books in all fifty states. They cover everything from siting waste management facilities to protecting sensitive environments. Those state regulations some-times involve local governments in their implementation, but for the most part standards are promulgated at the state level for statewide application and leave little room for local input or deviation from state rules. Some states, however, have supplemented direct state regulation with a collaborative approach to managing growth. This entails programs in which the states set policy goals and standards but leave, to varying degrees, the specific details of planning and land-use regulation to the discretion of local governments.

Florida is among ten states with programs that either require or strongly encourage local governments to adopt comprehensive plans and related land-use regulations.[1] Among those states, Florida is noteworthy for the degree of pre-scription and coercion it uses to secure adherence by local governments to state objectives. Florida's coercive policy provides a stark contrast to the cooperative intergovernmental mandates in New Zealand and New South Wales that we discuss in this book. Moreover, Florida's policy-makers have been moving in the direction of increased coercion and prescription in reforming environmental policy, which is the opposite of the movement toward increased flexibility of intergovernmental arrangements found in the other settings.

In this chapter, we trace the evolution of Florida's approach to environmental and growth management, noting adjustments state policy-makers made as they learned how to manage development and sought a clearer sense of what dimensions of sustainability to pursue. We begin with the array of single-purpose programs that policy-makers put in place in the 1970s aimed at protecting the environment in response to mounting citizen environmental activism. These are of interest because they illustrate state-dominated, prescriptive approaches to environmental management. These policies encountered serious obstacles because they failed to halt what many saw as continuous degradation in the quality of life. Our main focus in this chapter, and as a comparison point in this book, is Florida's revolutionary growth management legislation of the mid-1980s. With this policy, legislators took what for a state was a drastic step of threatening local governments with severe sanctions if they failed to plan and regulate growth in accordance with state standards. This chapter discusses the emergence of this policy, its key features, and the challenges for intergovernmental policy design and environmental management that the policy poses.

THE SETTING

Blessed with abundant natural resources and beauty, Florida long sought to promote migration to what, prior to 1950, was a sparsely settled, mostly rural southern state. Those efforts succeeded beyond the promoters' wildest dreams. From a population of less than three million in 1950, Florida grew by almost six million over the next 25 years, moving it from the twentieth to the eighth largest state. By the 1990 census, the population had surged past 13 million, and forecasters projected that number would double to 26 million by 2020.[2] Florida's population explosion created enormous wealth, but it also placed severe strains on the environment and physical infrastructure such as transportation and water and sewerage systems. Figure 2.1 depicts the Florida setting.

In addition, the rapid growth placed millions of people at risk from natural hazards. Florida contains a vast system of wetlands (they cover a third of the state), so that growth has been concentrated in low-lying areas near the coast. The Federal Emergency Management Agency estimated in 1991 that over 10,000 square miles in Florida are flood-prone, with over one million households and US$46 billion in property located in areas vulnerable to inundation in a 100-year storm.[3] Between 1982 and 1992, storms resulted in seven presidential disaster declarations, involving thirty-four of the state's sixty-seven counties. Hurricane winds may pose an even larger threat than flooding. When Hurricane Andrew came ashore in August, 1992, flooding caused little damage, but winds that gusted to almost 200 miles per hour produced some US$30 billion in losses.[4]

STATE EFFORTS TO ACHIEVE SUSTAINABLE ENVIRONMENTAL MANAGEMENT

By the mid-1960s, increasing vulnerability to floods and hurricanes, increasing pollution, increasing water shortages, and other environmental problems began to attract widespread public attention.[5] But Florida was ill-prepared to respond. The antiquated state government contained no means for coordinating action by some 200 independent and quasi-independent state departments and commissions. Local governments were equally hamstrung. One of the last states in the nation to authorize local land-use regulation, as the "environmental decade" of the 1970s dawned, less than a third of Florida's land area was covered by city or county zoning and subdivision regulations.

Like New Zealand in the 1980s, Florida in the 1960s first put its governmental

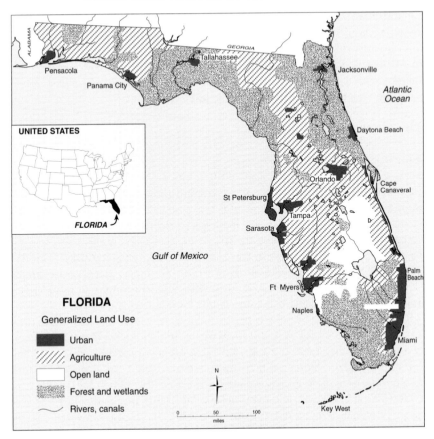

Figure 2.1 The Florida setting

house in order and then addressed issues of environmental sustainability. A new state constitution in 1967 consolidated state agencies into twenty-five departments, and legislation in 1972 vested flood control and other water resources functions in five water management districts, with boundaries set to match those of major watersheds. With reapportionment that reduced the representation of conservative rural areas in the state legislature, urban and urbanizing areas gained control of state policy-making for the first time. This began in the early 1970s.

Single-purpose policy mandates

Changes in the capabilities of state agencies and the balance of power among relevant interests set the stage for a revolution in state environmental policy. It began in the mid-1960s with efforts to halt the rampant destruction of shoreline ecosystems and gained momentum in 1971, when Governor Rueben Askew argued, "It's time we stopped viewing our environment through prisms of profit, politics, and geography or local personal pride," and convened the South Florida Water Management Conference to suggest a new course for state policy.[6]

The conference argued for a unified approach guided by the concept of limiting development to the carrying capacity of land and water resources. When carrying capacity proved difficult to translate into concrete policy, state legislators instead crafted a series of laws and programs, each focused on a visible consequence of growth pressures – cycles of flooding and drought, infrastructure strained beyond capacity, and destruction of critical environments – and called for a state plan to provide needed coordination. Each of these programs reflected the belief that local governments were not competent, on their own, to manage growth and its consequences. In some cases, state law makers preempted local authorities by creating new institutions to perform local functions, while in others they imposed new rules on management of development by local governments, threatening to impose direct state regulation if localities did not follow the state's directions. In the remainder of this section, the key features of these single-purpose laws are briefly recounted.[7]

Preserving beaches and shorelines Adopted in 1965, the *Beach and Shore Preservation Act* foreshadowed the outpouring of environmental programs that followed the South Florida Water Management Conference. In response to widespread destruction of frontal dunes by shorefront development projects and accelerated shoreline erosion and resulting property losses, the legislature created a state permitting program for construction below the mean high-water line of tidal waters. Local governments are required to notify the state Department of Natural Resources of permit applications covered by the legislation and to notify the applicant of the state's requirements. In 1971, the legislation was amended to also require local governments to impose additional and more stringent building standards seaward of a coastal construction control line, which is based on anticipated erosion rates and designed to include all the area subject to a 100-year storm surge from hurricanes. In addition to providing protection against coastal storms, the required coastal construction setback lines are designed to insure that development does not interfere with natural shoreline fluctuations or dune stability and recovery following storm events. While delegating administrative duties to local governments, the Department of Natural Resources retains the right of final approval when any deviation from its standards is proposed and the right to reassert administrative control, if required.

The *Beach and Shore Preservation Act* also created a trust fund, financed by annual appropriations from the state general fund, to finance up to 75 percent of the cost of beach restoration, beach renourishment, and hurricane protection. Projects are identified through a long-term, comprehensive state beach management planning process. In addition, the act gives the governor power to declare a shore erosion emergency, which can free up additional funds to alleviate erosion. Beach restoration projects also often qualify for federal funding of up to 50 percent of project costs under the *River and Harbor Act* of 1962. County beach and shore preservation districts are authorized to carry out local beach management planning and levy *ad valorem* taxes and issue bonds. The emphasis in all of these programs has been on the use of so-called "soft" shoreline protection methods, such as beach nourishment and sand dune restoration, rather than coastal armoring, through jetties, sea walls and groins, although the latter are authorized and have been used when soft methods are not feasible.

Controlling floods and drainage The *Water Resources Act* of 1972, one of a package of environmental programs enacted on the heels of the South Florida Water Management Conference, created the system of five regional water management districts. The districts operate under the loose supervision of the state Department of Environmental Regulation, but they have taxing powers and independent boards of governors appointed by the governor. The ample financial resources of the districts have allowed them to assemble large professional staffs to deal with state-mandated responsibilities that include flood control, drainage, and regulation of the consumptive use of water resources. They do not, however, have regional land-use planning authority, which instead has been given to regional planning councils whose boundaries do not coincide with natural boundaries. Thus, Florida failed to accomplish the integration of regional land-use planning with natural resource planning and management achieved in New Zealand, and instead vested these functions in separate agencies.

Protecting critical environmental areas The *Environmental Land and Water Management Act*, also enacted in 1972, established the Area of Critical State Concern program to protect areas deemed to have statewide environmental, historical, natural, archaeological, or other importance. The program provides two approaches to protect areas of state concern. With the first, local governments can volunteer to prepare a management plan by establishing a Resource Planning and Management Committee. If, however, a voluntary plan is not prepared or does not address the state's concerns, the second approach can be invoked with designation of an area of critical state concern by the governor and cabinet, sitting as the Administration Commission. Once designated, local governments are given six months to create land-use regulations that address state and regional concerns. The local effort can be rejected or amended by the state, and if the local government chooses not to act, the statute directs the state planning agency to adopt regulations.

Since 1972, only four areas have been designated by the state: Big Cypress Swamp, Green Swamp, and Florida Keys – all in south Florida, and Apalachicola Bay in northwest Florida. The program was slowed by a legal challenge (in which the Florida Supreme Court found that the delegation of legislative authority used to designate two of the areas was unconstitutional) and conflicts between economic development and environmental protection interests, which led the state to proceed with caution in designating areas. A special session of the legislature in 1978 re-established Green Swamp and the Florida Keys as areas of critical state concern. In addition, local governments proved reluctant to prepare the required plans and implementing regulations. For example, Monroe County, which contains the Florida Keys, refused to prepare a plan until 1982, when the state agreed to pay the required planning costs. Nevertheless, hurricane hazard mitigation has been fostered through this program. The Monroe County comprehensive plan uses evacuation time as the key limiting factor in determining how much new development can be accommodated in the Keys. Management plans prepared by eight locally established resource planning and management committees also address hurricane hazards and floodplain management. These plans cover large areas of Florida's coasts and the floodplains of important rivers.

Mitigating the effects of large-scale development A third component of the 1972 legislation addressed the region-wide impacts of large-scale projects, which at that time made up 5 to 10 percent of development taking place in Florida each year. The Developments of Regional Impact program, enacted as another part of the *Environmental Land and Water Management Act*, set up eleven regional planning agencies and required them to prepare reports and recommendations to local governments about the regional environmental, social, and economic impacts of proposed large projects. Those reviews include considerations related to public safety. Local governments are charged with examining whether proposed projects are consistent with local plans, local land-development regulations, and the recommendations of the regional planning council, and then making a final decision whether to approve the project. The state, regional planning councils, and developer have the right to appeal local government decisions to the governor and cabinet, sitting as the Land and Water Adjudicatory Commission.

Although the regional impact process drew praise from influential quarters, a number of shortcomings soon became evident. In particular, the role of regional councils proved difficult to implement, since local governments could ignore their suggestions and log-rolling among the local government representatives on the boards of the regional councils stymied efforts to assert regional environmental concerns. Impact assessment processes are usually aimed at mitigating adverse effects that spill over from projects to neighboring property and jurisdictions. But the regional reviews came late in the process of project planning, so that they did not serve that purpose well. Thomas Pelham, a law professor and former secretary of the Florida Department of Community Affairs, observed that the law was "seriously deficient in terms of substantive criteria for evaluation," and basically, "a process without policy."[8] Finally, the whole process drew continual fire from large-scale developers, who saw it as increasing project costs while providing few benefits to the public.

Instituting statewide building standards The *Florida Building Codes Act* of 1974 created the State Minimum Building Code. The code includes references to several nationally recognized model codes that may be relied on in meeting state requirements. It requires local governments to adhere to the requirements of the federal National Flood Insurance Program to elevate buildings to the predicted height of the 100-year flood. There were no special state requirements for building in coastal areas until enactment of the *Coastal Zone Protection Act* of 1985. The building codes act of 1974 imposes stricter building standards in specified areas to minimize damage to the environment, property, and life.

The emergence of comprehensive planning

Coordinated, comprehensive planning was the final ingredient in Florida's early efforts to gain control over environmental harms. The *Florida State Comprehensive Planning Act* of 1972 sought to strengthen planning and program coordination in state government by creating a Division of State Planning in the Department of Administration, which was to coordinate preparation of a state development plan. However, this program was "virtually stillborn."[9] The plan produced was too long and detailed and the goals and policies it contained too vague and contradictory for either the legislature to adopt or to serve as guides to land-use planning by other state, regional and local governments. The plan did not mandate action on its recommendations, it was merely advisory, and as a result, state and regional agencies generally ignored its suggestions.

Mandated local planning, consistent with state goals, was part of the initial draft of that 1972 legislation, but when that drew strong opposition from local governments, the idea was dropped. A study commission (Environmental Land Management Study Committee) appointed by Governor Askew to find ways to coordinate policy proposed a state law that would mandate local planning, but city and county governments again succeeded in defeating that idea in the 1974 session of the legislature. By 1975, however, continuing problems traceable to urban growth made the need for local land-use planning more evident than ever, and the legislature finally was able to override opposition by local governments and pass the *Local Government Comprehensive Planning Act* of 1975. This legislation required local governments to adopt comprehensive land-use plans and to adopt or amend land-use regulations then in force so that they were consistent with the comprehensive plan.

The state's comprehensive planning act was fairly prescriptive, but it lacked sufficient coercion or incentives to gain the attention of local governments. If local officials failed to plan within four years, the act authorized county governments or the state to plan for them. But, aside from an advisory regional and state review, no teeth were provided to ensure that plans addressed important state or regional goals or met minimal standards for plan quality, and no funds were provided to help local governments prepare the plans the state demanded. The legislation authorized US$50 million in planning grants to help local governments meet the costs of preparing the required comprehensive plans, but the funds were never appropriated. By 1980, five years after the enactment of the planning mandate, only about a fifth of Florida's local governments had prepared the required comprehensive plan and only about half had submitted at least one element of the plan to the state for review.[10]

The state planning mandate received little support from local government officials; the plans it spawned varied widely in quality; and without a means for the state to enforce internal consistency that the legislation required, local land development regulations and development orders frequently ignored plan prescriptions. Political scientist Lance deHaven-Smith summed up the system's failure, noting:

> Lacking clear state and regional policy, Florida's growth management system has evolved strong thumbs but weak fingers. . . . The system's weak fingers are at the local level in the planning, amendment, and review process. Plans exert very little influence over local land-use regulation. When there is a conflict between the plan and a desired zoning decision, it is often the plan rather than the zoning decision that is adjusted. Many local governments amend their comprehensive plans as often as every few weeks. . . . [required] regional and state review [are] simply circumvented.[11]

By the early 1980s, those deficiencies were so glaring that planning advocates mounted a second campaign to craft an effective state growth management program.

INTERGOVERNMENTAL COERCION AND PRESCRIPTION: DESIGNING A COMPREHENSIVE GROWTH MANAGEMENT PROGRAM

The failings of the single-purpose laws of the 1970s became evident in Florida as a variety of other growth management issues began to share center stage with environmental concerns. A study conducted by the University of Florida, for example, reported a backlog of US$10 billion in infrastructure for transportation, water supply, and wastewater treatment and the need for US$41 billion in additional public investment through the year 2000.[12] To cope with that crisis, local governments began shifting infrastructure costs to developers through the regional impact review process and a newly instituted system of impact fees in which developers were assessed charges to recover the off-site costs their projects imposed on public service systems. In combination, self-evident limitations of existing legislation, continued environmental degradation, citizen discontent with traffic congestion and other public service deficiencies, and developer discontent with the regional review process and the distribution of the costs of growth created the favorable political climate needed for sweeping changes in growth management legislation.

29

The new philosophy

In 1982, Governor Bob Graham formed a study committee, called Environmental Land Management Study Committee II, composed of a cross-section of interests concerned with growth management issues, and he told the committee to evaluate Florida's planning system. In its 1984 final report, the committee spelled out a new philosophy for growth management in Florida. It stressed the need for a system that embraced what we term the five Cs of modern growth management in the United States: consistency, concurrency, compactness, coastal management, and citizen participation. Between 1984 and 1986, each of the elements of the growth management philosophy of the Environmental Land Management Study Committee II was enacted into state law.

The first two are major innovations in planning arrangements by Florida policy-makers. Consistency embodies the notion that governmental policies at the state and local level should be in harmony. According to the ELMS II Committee, consistency could be attained if the state legislature put in place a top-down planning framework anchored by a legislatively adopted state plan. The overarching policy embodied in the plan would give the state a basis for reviewing regional and local plans to ensure vertical consistency among levels of government, while horizontal consistency at the local level would be secured through revitalized regional plans and by requiring local governments to pursue land-use regulation and public investments in a manner consistent with regional plans, plans of adjacent localities, and their own local plan. Concurrency is a concept similar to the ecological notion of carrying capacity that local governments should not authorize new development unless needed infrastructure capacity is already in place. The committee viewed this as a way to cope with strained infrastructure (development would be halted until needed capacity was in place) and to insure consistency between land use and capital improvement planning.

The remaining tenets of growth management are aspects of prior planning policies. Compactness was viewed as a way to limit urban sprawl and thereby protect agricultural land and sensitive environments. Coastal management was seen as essential "to reverse the practice of careless and reckless coastal development in high hazard areas susceptible to hurricanes and other storm hazards."[13] Citizen participation was seen as a way to ensure that local policy remains responsive to the interests of environmental and other local groups concerned with the course of growth and development. It could be achieved by requiring that citizens be given a role in the development of plans and by giving citizens standing to challenge plans and other local policy decisions when they felt the state's prescriptions were not being followed.

State, regional, and local planning

The *State and Regional Planning Act* of 1984 was the first legislative response to the committee's recommendations. The law mandated the preparation of a state comprehensive plan by the governor's office, which was to be presented to the legislature for adoption in 1985. It also required the preparation and adoption of state agency functional plans and regional comprehensive plans, each to be consistent with the overall state plan.[14] A first-time appropriation of US$500,000 to the state's regional planning councils supported the initial work on regional plans. That initial grant was supplemented by two million dollars in additional funds over the next two years, so that by the end of 1987 all eleven regional plans had been completed.

After holding extensive public hearings, the Governor's Office presented the proposed state plan to the legislature early in the 1985 session, and it was enacted as the *State Comprehensive Planning Act* of 1985 by a strong majority in both houses. The new state plan was reasonably concise but also comprehensive, containing twenty-seven goals and accompanying policies, including goals and policies dealing with public safety. Since its adoption, the plan has served as the cornerstone of Florida's growth management system.

The new integrated growth management system was finally put in place by passage of the second major law in 1985: *The Omnibus Growth Management Act*. One part of the omnibus law, the *Local Government Comprehensive Planning and Land Development Regulation Act*, mandated new local comprehensive plans and required that they be consistent with the goals of the state plan, the comprehensive regional policy plans, and other applicable statutes. It authorized the state to establish minimum criteria for local plans, which the Department of Community Affairs subsequently acted on through Rule 9J-5 of the Florida Administrative Code. This vertical consistency is complemented by the requirement for horizontal and internal consistency. Each local plan must include an intergovernmental element so that all local plans in a region are compatible with each other. Local plan elements also must be consistent with each other, a requirement that is important for hazard mitigation, since hurricane evacuation routes can be protected from overdevelopment by the requirement that the traffic circulation, coastal, and future land-use elements of local plans be coordinated and consistent with each other.

NATURAL HAZARDS UNDER THE NEW REGIME

The broad mandates of the growth management legislation are given shape and substance by Rule 9J-5, which sets minimum standards for judging the adequacy of local plans submitted for state approval. Rule 9J-5 requires certain elements in local plans and prescribes methods local governments must use in preparing plans (for example, Rule 9J-5 prescribes population projection methods to be used in plan preparation). Element by element, it lists the types of data, issues, and goals and objectives that must be addressed using almost a checklist format. It also requires specific, measurable objectives that can be used by the state to monitor the progress of local governments in meeting state goals.

Rule 9J-5 addresses natural hazards in the required coastal management element. Coastal management provisions of the plans contain several prescriptions relating to the efforts of local governments to address natural hazards:

1 limit public expenditures that subsidize development in high-hazard areas unless the expenditures are related to restoration and enhancement of natural resources;
2 direct population concentrations away from known or predicted high-hazard areas
3 maintain or reduce hurricane evacuation times; and
4 include post-disaster redevelopment plans to reduce exposure of human life and property to natural hazards.

These requirements signal a complete reversal in policy in a state that in the 1950s and 1960s was infamous for allowing wholesale subdivision and marketing of property that was not only flood-prone, but also often permanently under water!

In addition to the required coastal element, hurricane and natural disaster considerations are addressed in other components of the general comprehensive planning process. Floodplain management has to be considered in the future land use, conservation, and coastal management elements of comprehensive plans and also in land development regulations. Rule 9J-5 defined floodplains as areas that would be inundated during a 100-year flood event, the same standard used by the National Flood Insurance Program.

The 1985 *Omnibus Growth Management Act* amended the *Beach and Shoreline Preservation Act* to strengthen its provisions and coordinate its applications with local comprehensive plans. The amendments reemphasized the value of the beach–dune system as the first line of defense for the mainland against winter storms and hurricanes, the increased risks created by construction in these areas, and the considerable cost to the state associated with post-disaster redevelopment. The growth management amendments created a new "coastal building zone," larger than the area previously regulated by the *Beach and Shoreline Preservation Act*. Development within the zone is categorized as either minor (expendable in the event of a storm) or major. Habitable major structures must comply with the state building code, regulations of the National Flood Insurance Program, and be able to withstand at least 110 mile-per-hour winds. The zone includes all coastal barrier islands and the Florida Keys. A new "construction prohibition zone" was also created. This zone includes the area seaward of the predicted seasonal high water line for the next thirty years. The Department of Natural Resources cannot issue permits for development within this zone, and the restrictions on building must be disclosed when property is sold.

The growth management legislation ordered local governments to establish standards for the provision of community facilities and services and concurrency requirements to ensure that new development does not outstrip the capacity of available facilities. While the concurrency provision does not deal directly with hazards, it does control the extent to which overburdened, older, usually coastal areas of Florida can continue to grow. For example, an amendment to the Collier County comprehensive plan that proposed lowering the level of service standard for certain thoroughfares would have permitted a hurricane evacuation route to become too congested. Both the state and regional planning council objected, and Collier County eventually withdrew the proposal.

The growth management act prohibits state funding for bridges that connect previously unlinked barrier islands and any increase in local infrastructure capacity in ways inconsistent with local comprehensive plans. Finally, the act allows the state to eliminate local projects from any application for federal disaster assistance unless the local government has adopted or contractually obligated itself to adopt a hazard mitigation plan (before the disaster) addressing building codes, flooding, public infrastructure, public information systems, preventive planning measures to ameliorate potential storm damage, and the ability of certain land uses to locate or reconstruct in coastal high-hazard areas.

Plate 2.1 Damage to newer construction in Fort Walton Beach, Florida, in the aftermath of Hurricane Opal, 1995 (courtesy of Phil Flood, Florida Department of Environmental Protection)

THE IMPLEMENTATION DESIGN: REVIEWING PLANS AND COERCING COMPLIANCE

The Department of Community Affairs adopted a number of procedural rules for local government planning and state review of local plans for consistency with state policy. In combination with various sanctions that can be applied when local governments fail to follow state directives, the growth management law put in place a highly coercive program for securing local planning and growth management. Yet, policy-makers also added various local government capacity-building features as additional ways of inducing participation and increasing competence.

Prescription and coercion

In Florida the policy prescriptions and coercive elements of the program relate compliance of local governments with local comprehensive plan content and adherence to deadlines for plan completion. To back this up, an extensive system for review of local plans was established.

Substantive and procedural prescriptions With Rule 9J-12 Florida's planning administrators established an enforceable schedule of completion dates for all local plans. Amendments to adopted plans cannot occur more frequently than twice per year, and each amendment is subject to review by the state for consistency with state policy. At least once every seven years, local plans must be formally evaluated by the locality and then amended to put in place any necessary changes. These updates, termed Evaluation and Appraisal Reports, must incorporate changes in state and regional policy that have occurred during the interim period as well as respond to changes in community circumstances brought about by growth, decline, or other factors. The Department of Community Affairs is authorized to review the adequacy of the local plan update process, including required citizen participation, and to review proposed amendments to local plans to ensure that they are consistent with state and regional policy.

In Rule 9J-11, the state carefully specified the procedures all parties are to follow in preparing and amending comprehensive plans. The state process for reviewing local plans for compliance with state standards is highly prescriptive. It assures that all parties with a stake in the outcome of planning decisions have an opportunity to be heard. The growth management legislation specifically requires citizen participation in the process of preparing land-use plans, and it gives citizens the right to intervene if they believe that local planning or regulatory decisions contravene state policy. In addition, if a local government fails to adopt a required land-use regulation, the Department of Community Affairs may bring an enforcement action in the Florida courts to compel compliance.

Coercion The Florida legislature authorized sanctions for local governments that did not submit plans on time and for plans found not in compliance with the growth management act. The governor and cabinet, sitting as the Florida Administration Commission, established policies in 1989 for imposing sanctions. The Department of Community Affairs has the ultimate authority for determining the consistency of local plans with both state and regional requirements. The state can withhold 1/365 of state revenue-sharing funds for each day a local government's plan is late or held to be out of compliance. That sanction has been challenged, but it was upheld by the Florida courts. The sanction was augmented in 1990, when the legislature passed a five billion dollar transportation infrastructure funding package and made local government eligibility for transportation funds contingent on an approved local comprehensive plan. In 1993, the legislature added additional sanctions to spur local governments to prepare plan evaluation and review reports every seven years. After the deadline for the report passes, local governments cannot amend their plans, for example to allow for new development, until the state has reviewed and approved the process used in preparing the evaluation report.

Building capacity and inducing compliance

In addition to authorizing the stick (sanctions) to induce compliance, the Florida growth management program contains carrots (incentives) to build the capacity of local governments to plan for and manage development. The program has imposed significant costs on local governments.[15] Cities and counties spent US$130 million for comprehensive planning during the 1991–2 fiscal year compared to US$55 million in the 1984–5 fiscal year, when the second round of growth management legislation was enacted. Per capita expenditures for planning increased from US$5.47 in 1984–5 to US$9.55 in 1991–2. In the first decade of the program, Florida governments will spend approximately one billion dollars on planning. Various state financing programs have helped lower costs of compliance for local government while also helping reduce political opposition and local efforts to repeal the program.

Financial assistance   Florida has provided significant financial assistance to help local governments prepare comprehensive plans and bring land use and other regulations into conformity with local plans and state growth management policies. Between 1985 and 1993, the state provided local governments with a total of US$36 million in planning grants, approximately four percent of the US$909 million in total local planning costs over that period.[16] In fiscal year 1994, no new grants were made, since by then all of the required local comprehensive plans had been prepared, but the governor requested an additional US$2.5 million to cover new costs imposed by legislation in 1993 that requires local governments to revise their comprehensive plans so that they are consistent with state policy.

Technical assistance   The Department of Community Affairs has worked to improve the capacity of local governments to plan for and manage development by formulating and disseminating model comprehensive plan elements and a model land development code for a mythical Florida city. The conservation and future land-use elements address planning for floodplains, and the coastal management element includes a plan for hurricane evacuation. The department disseminates two additional technical aids: a newsletter with advice on compliance and copies of plan evaluations that highlight sections of plans the department rejected because they were not in compliance with Rule 9J-5. Technical assistance provided to local governments by the Department of Community Affairs is supplemented by technical assistance from the eleven regional planning agencies.

COMPARISON OF THE 1965–84 AND 1985–94 PLANNING REGIMES

The radical changes brought about by the growth management laws enacted in the mid-1980s and the newer regime's more prescriptive and coercive character is summarized in Table 2.1. In place of a generally weak and largely cooperative planning system from 1965 through 1984, Florida put in place in 1985 a very strong, highly prescriptive and coercive program to both plan for and manage urban development in ways consistent with state policy objectives. This was accompanied by a number of other reforms in Florida's environmental management legislation.

Between 1965 and 1984, Florida put in place many of the ingredients needed to accomplish key objectives for environmental management. The emphasis on restricting development in environmentally sensitive areas made the goal of what we now call environmental sustainability a centerpiece of these policies. The emphasis was on the preservation of beaches and other critical environments for the future. The quality of the environment and protection from natural disasters was advanced by state initiatives in shoreline and water resources planning, public investment in flood protection, and establishment of uniform building code requirements. The Developments of Regional Impact regulations established procedures to consider and then ameliorate the region-wide effects of large-scale projects. The critical areas programs protected some of the state's most valuable resources. But a key ingredient of statewide, systematic, coordinated land-use planning and management was missing. Without sufficient incentives or sanctions to foster effective local planning, sprawling, resource-consuming urban growth continued unabated.

The environmentally-focused planning regime that lasted a decade beginning in the mid-1970s contained many of the pieces needed for growth management, but the state's planning mandate was sorely deficient. It had some procedural prescription (e.g., deadlines for local plan preparation), but no incentives (e.g., financial assistance) and weak sanctions to help build local commitment to planning. While some planning took place where environmental interests had gained power, many local governments continued indiscriminately to foster growth while generally ignoring its adverse consequences. To many observers,

36

Table 2.1 Florida's growth management regimes

Components	Growth management regime (1965–84)	Growth management regime (1985 to present)
Intergovernmental policy	Prescriptive but non-coercive	Prescriptive and coercive
Sustainability concept	Aspects of environmental sustainability	Aspects of sustainable development
Policy prescriptions		
Procedural	Restrictive and limited	Restrictive and extensive
Substantive	Flexible and limited	Restrictive and extensive
Sanctions (*vis-à-vis* local governments)	Weak	Strong
Incentives		
Grants and subsidies	Low	High
Technical assistance	Low	High

however, the principal flaw of the old system was lack of substantive prescription. Without a state plan, there was little basis for state review of local plans or policy guidance to regional councils in their review of large-scale development projects.

The reforms enacted between 1984 and 1986 took the discretion to plan and manage urban growth away from local governments and literally forced them to both plan and act in ways consistent with state goals. The comprehensive planning reforms corrected perceived deficiencies in environmental and growth management by putting in place a state plan to provide policy guidance and a long list of detailed specifications for the content of local plans. Where the previous regime paid little attention to building local capacity to plan and manage growth, the new system provided a number of grant-in-aid programs to foster local planning and a large staff in the Department of Community Affairs to provide technical assistance. Along the way, the emphasis shifted somewhat from environmental protection to a broader planning regime emphasizing both environmental and social concerns. In so doing, the policy emphasis changed from that of environmental sustainability to incorporation of the broader concepts of sustainable development.

ISSUES RAISED BY THE FLORIDA GROWTH MANAGEMENT PROGRAM

In the ten years since the Florida legislature enacted the new growth management system, a number of concerns have been raised by various commentators. We conclude this chapter by briefly recounting those issues most relevant to the themes of this book.

The appropriate degree of coercion and prescription

The top-down nature of Florida's growth management system and extreme degree of prescription and coercion have been controversial. Local governments have objected vehemently to state intervention in local land-use matters, and landowners have complained of excessive governmental intrusion on private property rights. One legal analyst describes these concerns in writing: "The implementation of the local planning program, particularly with the strong state role exercised by the Department of Community Affairs, has heightened concern among landowners and developers about a diminution in their ability to use private property."[17] Responding to these concerns, a bill to scuttle the whole program was introduced in the Florida legislature in 1992 and drew some support, although it eventually failed to garner the votes needed for passage.

Nevertheless, the governor took note and appointed a new Environmental Land Management Study Committee (ELMS III) in 1992 to find ways to make the whole system more "user friendly" by simplifying administrative processes. That led to a major overhaul of the system in 1993 to improve the efficiency of state review of local planning. In that legislation, the state legislature also reiterated the need for just compensation for any regulatory taking of private property and urged local governments to be sensitive to private property rights in local planning and to "not be unduly restrictive" in implementing the local planning program.

A related issue arose over the degree of specificity in state standards. According to an early review of state growth management systems by legal educator Daniel Mandelker, a key question raised by these laws is: "Can goals and guidelines be made sufficiently precise and substantive so that they can provide an adequate basis for the exercise of state powers of local review?"[18] In 1992, the committee proposed the state deal with that issue by adopting a two-tiered state plan. One tier would contain general policy goals, and a second tier (a proposed Strategic Growth and Development Plan) would provide clearer policy guidance to state, regional, and local agencies. That proposed approach drew stiff opposition from local governments, however, because it would have increased the prescriptive nature of the state program in ways that could not easily be forecast. Reflecting continued controversy surrounding the role of the state plan in guiding local land-use planning and decision making, in the 1994 session of the Florida legislature a bill that would have significantly weakened the state plan passed both houses, but was vetoed by the governor.

The issues of state intrusion on local government authority and diminution in value of private property in Florida have been played out in other state growth management programs in the United States, as well as elsewhere, as the experience of New South Wales with its initially coercive floodplain management policy reveals. In commenting about the governance issues William Fulton suggests, in fact, that this may be a constant factor in such programs: "Local governments can only be expected to try to keep land-use decisions in their own hands."[19] As a result, according to planning educator Dennis Gale, " . . . the trend appears to be away from the state dominant model . . . " and toward paradigms that invest more power with local governments.[20] While that may be true, Florida, with its highly centralized system, has yet to significantly weaken state control, in spite of complaints by local governments. That suggests that a coercive intergovernmental approach can survive politically, particularly when there is enough slack in the system.

The role for regional planning councils

Regional planning councils were given five important roles in the Florida growth management system: policy planning at the regional scale; review and coordination of local government plans; resolution of disputes among local governments; technical assistance to local governments; and permit review for large-scale development projects. But, from their inception in 1972 the regional councils have been politically controversial and, according to some commentators, not very effective.[21] Weaknesses include lack of resources from the state to perform their mandated functions (for example, the eleven regional councils had an aggregate budget of just US$1.55 million in the 1990–1 biennium for review of local plans and technical assistance) and an inability to coordinate development decisions with the decisions of other regional agencies, such as the five water management districts.[22] In addition, even though local officials hold two-thirds of the seats on each regional council (the other third are appointed by the governor), local governments resent regional interference in local affairs, just as they resent state interference.

Reflecting the political unpopularity of regional regulation, the 1992 session of the legislature acted to repeal the law establishing regional planning councils (to "sunset" them) if the 1993 legislature did not re-enact the basic enabling legislation. In 1993, the legislature retained the regional councils but eliminated their role in reviewing developments of regional impact and circumscribed regional planning by forbidding councils from adopting standards for public services that would affect local provision of services or development decisions based on the adequacy of services. At the same time, however, the legislature expanded their role in regional planning by mandating that they establish dispute resolution processes, establish a cross-acceptance process (as an alternative to litigation) to resolve disputes among local governments and increase coordination among regional and local plans, and expand their review of proposed amendments to local government plans.

The difficulty in crafting a regional planning program in Florida reflects broader difficulties in regional planning in state growth management programs in the United States.[23] A number of states have provided modest roles for regional agencies in growth management, usually involving coordination and technical assistance responsibilities, but only Vermont gave regional planning agencies the power to approve local plans, and that power was delayed indefinitely in 1990 due to protests over it.[24] The issue of appropriate regional roles is also pervasive in our discussion in Chapter 3 of New Zealand's governmental and environmental management reforms and in our discussion in Chapter 6 of the precarious role of regional entities in environmental management.

The bottom line: how effective is the Florida system?

The bottom line continues to be the ability of intergovernmental policy to affect the behavior of local governments in managing development and land use. The report card on Florida growth management system was still being written when the empirical work for this book was undertaken. However, some mid-term grades already have been given that raise doubts about the Florida program's efficacy. We further examine the program's performance in Part II of the book.

NOTES

1 Other states with statewide comprehensive growth management systems include: Georgia, Hawaii, Maine, Maryland, New Jersey, Oregon, Rhode Island, Vermont, and Washington.

2 Figures cited from Allan D. Wallis, "Growth Management in Florida," Working Papers Series (Cambridge, MA: Lincoln Institute of Land Policy, 1993); and John M. DeGrove, "The Politics of Planning a Growth Management System: The Key Ingredients for Success," *Carolina Planning* 16, no. 1 (Spring 1990): 26–34.

3 L.R. Johnston and Associates, *Floodplain Management in the United States: An Assessment Report*, Volume 2: Full Report (Washington, D.C.: U.S. Government Printing Office, 1992). A 100-year storm is an event that has a 1 percent chance of occurring in any given year. Hazard specialists refer to this as a 1-in-100 year event.

4 Loss data cited from Federal Emergency Management Agency, *Interagency Hazard Mitigation Team, Interagency Hazard Mitigation Team Report in Response to the August 24, 1992 Disaster Declaration for the State of Florida*, FEMA-955-DR-FL, Hurricane Andrew (Atlanta, GA: Federal Emergency Management Agency, 1992).

5 The environmental problems that led to state intervention are chronicled in William R. McCluney, *The Environmental Destruction of South Florida* (Miami, FL: University of Miami Press, 1971), and Raymond Dasmann, *No Further Retreat* (New York: Macmillan, 1971).

6 Quoted in Luther Carter, *The Florida Experience: Land and Water Policy in a Growth State* (Baltimore, MD: The Johns Hopkins University Press for Resources for the Future, 1975), p. 126.

7 Not addressed here are a plethora of state pollution control laws and programs that stem primarily from efforts of the federal government to abate water pollution, air pollution, and the adverse effects of exposure to hazardous wastes. For a review of the states' role in pollution control in the United States see Evan J. Ringquist, *Environmental Protection at the State Level: Politics and Progress in Controlling Pollution* (Armonk, NY: M. E. Sharpe, 1993).

8 Thomas G. Pelham, *State Land-Use Planning and Regulation: Florida, the Model Code, and Beyond* (Lexington, MA: Lexington Books, D.C. Heath, 1980), p. 193.

9 Comments by Allan D. Wallis, "Growth Management in Florida," op. cit., p. 5. Also see: John M. DeGrove, *Land Growth and Politics* (Chicago, IL: APA Planners Press, 1984) and Thomas G. Pelham, *State Land-Use Planning and Regulation*, op. cit.

10 Edith Netter and John Vranicar, *Linking Plans and Regulation: Local Response to Consistency Laws in California and Florida*, Planning Advisory Service Report No. 363 (Chicago, IL: American Planning Association, 1980), p. 12.

11 Lance deHaven-Smith, "Regulatory Theory and State Land-Use Regulation: Implications from Florida's Experience with Growth Management," *Public Administration Review*, 44, no. 5 (October, 1984): 413–420.

12 Reported in "Infrastructure: Keeping Pace with Growth and Change," in *The Saddlebrook Papers: A Reader on Growth Management in Florida*, prepared by the Florida House Select Committee on Growth Management, 1984 as cited in Allan D. Wallis, "Growth Management in Florida," op. cit., p. 7.

13 John M. DeGrove with Deborah A. Miness, *Planning and Growth Management in the States: The New Frontier for Land Policy* (Cambridge, MA: Lincoln Institute of Land Policy, 1992), p. 14.

14 In 1993, the legislation was amended to focus the attention of regional plans on strategic regional issues and ensure that they did not intrude on purely local matters. Now regional plans are required to address only five topics: affordable housing; economic development; emergency preparedness; natural resources of regional significance; and regional transportation. Councils can, at their discretion, deal with other subjects of regional importance.

15 The following figures are from Florida Advisory Council on Intergovernmental Relations, *Fiscal Impact of Comprehensive Planning Requirements on Florida's Counties and Municipalities – with an Emphasis on the 1993 ELMS Legislation* (Tallahassee, FL: The Council, January 1994).

16 "Paying the Piper for Planning in Florida," *The Growth Management Reporter* 2, no. 1 (1994): 1–2. For an economic analysis of the significant costs imposed on local governments by Florida's growth management legislation, see K.T. Liou and Todd J. Dicker, "The Effect of the Growth Management Act on Local Comprehensive Planning Expenditures: The South Florida Experience," *Public Administration Review* 54, no. 3 (May/June, 1994): 239–244.

17 David L. Powell, "Managing Florida's Growth: The Next Generation," *Florida State University Law Review* 21, no. 2 (1993): 223–340.

18 Daniel R. Mandelker, part of a longer discussion by Mandelker in Scott A. Bollens, "Restructuring Land Use Governance," *Journal of Planning Literature* 7, no. 3 (February, 1993): 211–226.

19 William Fulton, "Land Use Planning, A Second Revolution Shifts Control to the States," *Governing* 2, no. 6 (1989): 40–45.

20 Dennis Gale, "Eight State-Sponsored Growth Management Programs: A Comparative Analysis," *Journal of the American Planning Association* 58, no. 4 (Autumn 1992): 425–439.

21 Thomas G. Pelham, "The Florida Experience: Creating a State, Regional, and Local Comprehensive Planning Process," in *State and Regional Comprehensive Planning: Implementing New Methods for Growth Management*, ed. Peter A. Buchsbaum and Larry J. Smith (Chicago: Section of Urban, State and Local Government Law, American Bar Association, 1993), pp. 95–124.

22 See John M. DeGrove and Deborah A. Miness, *Planning and Growth Management in the States: The New Frontier for Land Policy*, op. cit.

23 See John M. DeGrove and Patricia M. Metzger, "Growth Management and the Integrated Roles of State, Regional, and Local Governments," in *Growth*

Management: The Planning Challenge of the 1990s, ed. Jay M. Stein (Beverly Hills, CA: Sage Publications, 1993), pp. 3–17.

24 Judith E. Innes, "Implementing State Growth Management in the United States," in *Growth Management: The Planning Challenge of the 1990s*, ed. Jay M. Stein (Beverly Hills, CA: Sage Publications, 1993), pp. 18–43.

3

DEVOLUTION AND COOPERATION

Resource management in New Zealand

In 1984, the newly elected Labour Government of New Zealand embarked on the comprehensive reform of government structures and roles. The reforms were described at the time by a New Zealand political historian as being "the greatest changes anywhere, anytime, in any democracy."[1] The reformers sought to achieve several fundamental changes. One was to get government out of business while bringing business into government. Another was to get central government out of local problems, while increasing the attention of local governments to more than parochial concerns. A third change was to remove cradle-to-grave dependency of citizens on the state and replace it with a "user-pays" self-reliance in a free-market system. Environmental planning and management were important elements in the overall reform process, and these changes have also been acclaimed internationally as being at the cutting-edge of sustainable management of natural and physical resources.

In this chapter, the reforms in New Zealand's management of natural resources and the environment are examined within the context of the broader set of radical governmental and social reforms. The intergovernmental roles and responsibilities that emerged from the reforms of central, regional, and local government and of resource management law are emphasized. Because it is central to the environmental and administrative reforms, a major part of the chapter addresses the policy intent and implementation of the *Resource Management Act* of 1991 and its implications for managing the environment. Finally, some reflections are offered on the challenges for implementation and administration that are posed by the new environmental-management regime.

THE SETTING

The extent of reform is remarkable for a nation of low-density population and a history of strong central government. As shown in Figure 3.1, the country's population of 3.5 million people is concentrated mostly in six urban areas each with a population exceeding 100,000. The total land area is similar in size to that of the American state of Oregon, but the separation of the country into a

43

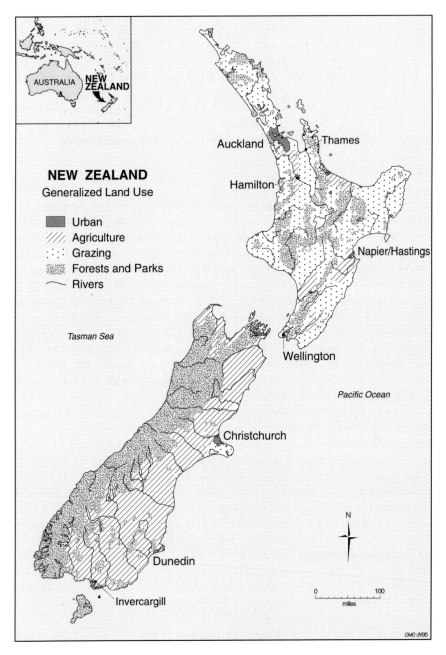

NEW ZEALAND
Generalized Land Use

- Urban
- Agriculture
- Grazing
- Forests and Parks
- Rivers

Figure 3.1 The New Zealand setting

North and South island creates greater isolation than the land area alone would suggest. New Zealand confronts a diverse set of natural hazards for which serious flooding occurs on average every other year, while cyclones strike on average twice every decade. Severe wind, flooding, landslides, and coastal erosion accompanied Cyclone Bola in March 1988, causing losses exceeding NZ$380 million. The previous year a serious earthquake caused a similar magnitude of losses in the rural Bay of Plenty on the North Island. An earthquake event of that size in Wellington would likely cause NZ$10 billion in losses, despite two decades of efforts to avert earthquake losses in the nation's capital.

THERE'S GOT TO BE A BETTER WAY

For a century after provincial government was disestablished in New Zealand, there existed an uneasy relationship between central and local government, particularly following pluralistic governance that emerged out of the 1930s depression.[2] Local government officials often complained about the power and interventionism of central government, yet efforts by central government to guide territorial local authorities (municipalities and counties, henceforth referred to as local councils) into better planning often met stiff resistance. Central government relied on inducements such as technical assistance and/or financial subsidies to direct local government planning, particularly for public works (e.g., sewerage systems, flood control schemes, and transportation). But the power of local government explains to a large extent why the many prescriptive mandates for environmental planning that emerged over the decades to 1984 did not contain strong coercive features. It also helps explain why central government had great difficulty introducing a tier of regional government to help facilitate its planning policies.

Resource development control

As the manager of much of the country's resources, central government became the dominant developer, especially after the Second World War. A great deal of public and private development was financially subsidized by the state. This often resulted in over-use and degradation of the resource base resulting from excessive applications of fertilizer on farmlands, clearing forests from marginal hill-lands, overgrazing land susceptible to erosion, and recurring floods and drought. Moreover, the New Zealand government was one of the biggest culprits in not complying with local government environmental policies and plans.

As environmental problems grew in the 1960s, public pressure was exerted on central government for developers to take account of environmental values. Policy-makers responded piecemeal by developing new legislation and/or amending the old.[3] The consequence was that by the time libertarian reforms were undertaken in the mid-1980s, environmental planning had spread across some seventy statutes that variously involved twenty central government

45

ministries and departments, twenty regionally based catchment authorities, twenty-two united and regional councils, 234 local councils, and a multitude of other *ad hoc* authorities like harbor boards and pest destruction boards. Not surprisingly, planning for resource development and environmental protection was therefore unduly complex and often contradictory.[4] Many informed people agreed that the planning system warranted reform, although opinion differed over why and how this should be done.

Controlling development in hazard-prone areas

The problems that central government experienced in having its policies translated into local practice are illustrated by efforts to control development in hazard-prone areas. The opportunity to guide growth and development locally and regionally through regulatory planning schemes was first offered by central government to local councils in the mid-1920s.[5] Because progress was minimal, preparing land-use plans became a requirement in the 1950s. Although the *Town and Country Planning Act 1953* was quite strongly prescriptive (procedurally and substantively), penalties for non-compliance by local councils were almost non-existent. For example, more than twenty years after it became law in 1953 for local councils to have operative district scheme plans for regulating land use, 50 percent of them still did not have one.[6] United (regional) councils remained very weak some ten years after they were mandated by central government in 1978 to undertake regional planning and civil defense functions, because local councils had political and financial control over them.

As part of a wide range of other land-use planning functions, regulations in 1960 required local councils to map and disclose natural hazards in their district plans and "provide as far as practicable, against land being used for purposes for which it [was] not suitable, having regard to . . . flooding, erosion, and land-slip . . . "[7] Councils also had to consider the danger of land being eroded or inundated by sea, river, or lake when approving subdivisions. These requirements were, however, readily compromised by conditional clauses, like "normally," "in council's view," and "unless otherwise mitigated." As well, lawyers acting for local councils often cautioned against public disclosure of hazards lest property owners take legal action. This was in spite of central government repeatedly warning that, providing decisions by local councils were based on the best available technical advice, the reverse situation applied. Since prescribing what ought to be done was not welcomed by local politicians and property developers, very little progress was made on controlling development in hazard-prone areas over the next two decades.

Although the legislation was progressively tightened in the late 1970s, a survey in 1983 showed that less than one-third of 103 flood-prone communities had flood maps in their district plans.[8] Where maps did appear most were inadequate for planning purposes. As a consequence, district schemes continued to contain limited land-use controls for natural hazards. Indeed, by the time the reform process began in 1984, developers and local councils had successfully blunted the legislative gains for regulating subdivision and building in hazard-prone areas by successfully lobbying central government to make favorable amendments to the relevant planning acts.

Local councils had little to fear from central government for not complying with its regulatory prescriptions. And, for their part, developers could expect minimal penalties for violating development consents, making prosecution a costly option for any regulatory authority. Similar problems arose in controlling land use for soil conservation and water quality purposes. For example, the maximum fine for unauthorized spillage of wastes into river systems in the 1970s was only NZ$3,000.

Central government made regional catchment boards a voluntary matter under the *Soil Conservation and Rivers Control Act 1941*. When in the first twenty years tax-payers in only two-thirds of New Zealand had opted for catchment boards, central government made them mandatory under the *Water and Soil Conservation Act 1967*. This act also created regional water boards within the catchment board system. Their main function was to issue water rights in order to control the allocation and quality of water, and to underscore the flood control and land drainage functions of the previous act. Combined, these acts had the primary aim of encouraging local councils to buy into the construction of land treatment and flood protection works, including extensive central government grants. Since similar subsidies were not available to local councils for controlling development in hazard-prone areas through land-use and building regulations, the playing field for managing flood hazards was far from level.

Clearly, without strong sanctions, any planning regulations could be easily compromised by local councils, especially with regard to development controls and natural hazards. As a future Minister of Finance was wont to say about New Zealand's major problems: "There's got to be a better way!"[9]

ENVIRONMENTAL ADMINISTRATION AND RESOURCE MANAGEMENT LAW REFORMS

In New Zealand history, 1984 is a watershed year. A crippled economy, rampant government interventionism, and state-sponsored environmental vandalism faced the newly elected Fourth Labour Government. Ironically, it set about dismantling the welfare state created by the First Labour Government in the 1930s.[10] The reforms were pervasive, radical, and fast moving. They were under-pinned by New Right ideology of monetarism, market liberalism, and public sector reform. Changes in environmental administration, policy, and planning have to be seen in the context of this overall reform process which also aimed at reducing the size and scale of central government and decentralizing its decision-making functions to regional and local levels.

The set of reforms enacted between 1985 and 1991 included the consideration of economic instruments in policy implementation; commercialization of the public sector; corporatization of the commercial operations of government (a prelude to privatization of state assets); reorganization of national environmental administration; restructuring of regional and local government; and integration of resources and environmental management statutes. The last three reforms relate directly to environmental policy-making and planning, but were pervasively affected by the ideology underpinning the first three.[11]

Restructuring government

Before reforming regional and local government, the departments and ministries of central government were restructured. Of particular significance for environmental policy and planning was the development of a new central government structure for administering environmental policies and operations.

National environmental administration Prior to reform, most resource development agencies of central government had multiple and often conflicting functions (e.g., resource development versus conservation). In its first term of office from 1984 to 1987, the Labour Government disestablished and restructured them into single-function agencies dealing with either policy, conservation, development, provision of services, or research. The environmental protection and enhancement elements from these various multi-functioned agencies (such as the New Zealand Forest Service, Department of Lands and Survey, Ministry of Works, and Department of Internal Affairs) went into three new agencies: a Department of Conservation charged with conservation advocacy and operations; a Ministry for the Environment focusing on environmental policy; and the Parliamentary Commissioner for the Environment established as an environmental watchdog. These three agencies formed the new environmental administration within central government.[12]

In keeping with a down-sized public sector, the new environmental administration is very lean. By 1993, the Ministry for the Environment had only 100 full-time equivalent staff, about seventeen of whom worked in three regional offices directly servicing regional and local government, and resource user groups. The agency's budget in 1994 was NZ$22 million of which some 45 percent was allocated as subsidies or other forms of grants to regional and local governments. The Ministry was charged with providing environmental policy advice, monitoring implementation of the *Resource Management Act* and its effects, and facilitating environmental management at all levels of government. For its part, the Department of Conservation is organized into fourteen regional conservancies responsible to a head office. Constant restructuring resulted in a decline of permanent staff – in the two years to 1993 – by 20 percent to 1,355. For the same period the annual budget fell 10 percent to NZ$117 million to cover not only its coastal responsibilities and statutory planning requirements under the *Resource Management Act*, but also the operational activities involved in managing the Crown estate. The Parliamentary Commissioner's office has 20 staff, including clerical workers, and an annual budget of NZ$0.6 million to carry out its auditing and watchdog role.

Local government reform In keeping with the philosophy of devolution and the view that local communities should take greater responsibility for the effects of their resource development decisions, policy-makers set about restructuring and reforming regionally based authorities and local councils after the Labour Party was re-elected in 1987.[13] An amendment to the *Local Government Act* in 1989 amalgamated the existing forty-two united/regional councils and catchment/regional water boards into thirteen new regional councils based on major catchment boundaries containing on average 235,000 people, the minimum being 50,000. These new regional councils absorbed most of the devolved functions of their predecessors, along with those of the pest destruction board, noxious plants authorities, and numerous land drainage and river boards. Under the *Resource Management Act*, they were made responsible for deciding on policies, objectives, and/or plans for integrated management of natural resources in their regions, and were given control of water, pollution, soil erosion, natural hazards, and the coastal marine area. Regional councils also have other responsibilities such as coordinating the planning for transport and civil defense in their regions.

The existing 234 local councils (i.e., territorial local authorities composed of towns, boroughs, cities and counties) were amalgamated into seventy-three district and city councils (henceforth called local councils), containing on average 40,000 citizens, the minimum being 10,000. These new local councils took over such functions of the previous local councils as managing water supply, sewage and waste disposal, roads, parks, and reserves, land subdivision, land-use planning and community development, along with health and building inspection, organizing civil defense, pensioner housing and libraries. With regard to the *Resource Management Act*, the new local councils are responsible for developing policies and/or plans on the use of land resources and for controlling the effects of land use, subdivision of land, hazards, and noise in their territories.

The territory of each new regional council contained several newly created local councils. However, local councils are authorized to take on the dual functions of regional and local councils as unitary authorities, an option strengthened by the central government in an amendment to the *Local Government Act* in 1992. Thus, there are now sixteen regional councils (which include four unitary authorities) and seventy-three local councils consisting of fifteen city councils and fifty-eight district councils (which include the four unitary authorities).

The regional and local councils came into being following elections in November 1989. Both types of council have elected members or councillors. The numbers vary according to the population in each constituency. An amendment to the *Local Government Act* in 1992 prevented a person being a councillor on both a regional and local council. Both councils are financed through a property tax or rating system. While local councils strike a general rate to cover all their services, regional councils strike both general and special rates such as for property owners who benefit from flood-control works. Both regional and local councils can delegate functions under certain circumstances.

To the new regional and local councils were applied the New Right principles already underpinning the reform of central government agencies. These included the separation of policy and regulatory functions from operational functions, performance-based contracts, objective-based planning, transparency of decision-making, and improved accountability. In commenting about the relevance of local government reform for environmental policy, the scholars Bührs and Bartlett emphasize that: "unquestionably local government reform's greatest significance for environmental policy was in establishing the structures, organisations, and basic processes that would be used by the *Resource Management Act 1991* to assign functions and redesign policy processes and institutions."[14]

Resource management law reform

Running parallel with the reform of local government in 1987 to 1990 was the review of resource management law. A long-standing criticism of the statutory planning process by developers and politicians was that meeting requirements for multiple resource consents (i.e., permits for various development activities) from different agencies under various statutes was too demanding, time-consuming, and costly. Critics thought that the planning process was over-regulated and unduly focused on the control of activities; in other words, it was procedurally and substantively too prescriptive. In addition, these groups asserted that planning had become unnecessarily complex and inflexible and therefore was an impediment to development, growth, and jobs.[15]

At another level, the national government had shown reluctance to be bound by the statutory consents process for its resource development projects, and thereby the environmental protection and enhancement procedures developed by the Commission for the Environment which it had established through a cabinet (executive) decision in 1972. When seen to be impeding implementation of its "Think Big" resource development projects in the late 1970s, the central government pushed through the highly controversial *National Development Act 1979* to fast-track the planning process for projects deemed to be in the national interest. For their part, environmentalists and other like-minded people wanted a system of environmental planning that would not be so readily compromised by public and private developers. The Minister for the Environment put it this way when reporting on progress with the resource management law reform process at the final conference of catchment authorities:

> The Resource Management and the Local Government Law Reform
> at present underway are not just a change for change's sake. Anyone in

doubt about that should glance through a few of the 3000-plus public submissions we've had on the subject. They will see that this reform is widely perceived as being badly needed and long overdue. It is the Government's intention to respond to that call. . . . [I]t wants objectives for both the process of decision making and the outcomes of those decisions for the environment.[16]

Although streamlining the overall planning process became a major objective of local government and resource management law reforms after 1987, calls for rationalizing and integrating resource management had been made long before. Starting in the mid-1970s, intergovernmental committees had been established to review legislation and structures pertaining to the management of the coastal zone, soil and water resources, and town and country planning. Also under review was the conflict between developing and conserving natural resources, and the need to strike some balance based on the then emerging principles of sustainable management. About that time, a review by the United Nations Organization of Economic and Cultural Development called for sweeping changes to the environmental administration of New Zealand's central government to avoid having multiple functions within one agency.[17] During this time there were emerging and noteworthy problems relating to environmental quality, particularly for pollution and hazardous substances. These fell through the cracks of piecemeal administrative structures. The various reviews were eventually subsumed in the comprehensive review of all resource legislation aimed at integrating them into a single piece of legislation.

The *Resource Management Act 1991* replaced fifty-nine statutes and nineteen regulations and orders pertaining to environmental planning. This was to be the sixth and final element in the reform program of the Labour Government. It was carried out by the Ministry for the Environment using a wide-ranging process of expert and public consultation. The pervasive presence of the Department of Treasury in the process was to ensure that the outcomes would be consistent with the overall philosophy of economic reforms. The local government and resource management law reforms were tightly coordinated through one cabinet committee headed by the Minister for the Environment, who was also Prime Minister and keen for success.

THE RESOURCE MANAGEMENT ACT: POLICY INTENTIONS AND IMPLEMENTATION DESIGN

In this section, we begin by providing an overview of the intentions that lie behind the *Resource Management Act* and then explain its implementation design – how it is supposed to work in practice. The legislation is complex and far-reaching. Although it builds on the previous legislation, there are six important changes from the previous planning regime, the first three of which are critical for new ways of doing business. First, and most important, the main goal is to enhance the environment through sustainable management of natural and

51

physical resources. Second, to achieve this goal, the legislation is driven through outcomes-based environmental assessments, rather than through activities-based prescriptions. Third, and most controversial in terms of intergovernmental relationships, the new regional councils were given the role of strategically managing the natural resources in their regions in an integrated fashion. Fourth, the legislation provides for integrated pollution management, makes better links between land and water management than before, and treats mining like any other land use. Fifth, there is more opportunity for public participation in decision-making, including better provisions for resolving conflicts through a pre-hearing process. Concomitantly, there are stronger enforcement measures authorized for use in securing compliance of the private sector with locally developed rules. Finally, there is greater attention to the *Treaty of Waitangi* and Maori resource management issues.[18]

Policy intentions

The *Resource Management Act* provides an overarching framework for integrated and sustainable management of natural and physical resources. The main components include the assessment of environmental effects (at a policy and operational level); policy analysis and plan preparation; decision-making (by local government politicians and the Planning Tribunal); and monitoring environmental quality through periodic reviews of policies, plans, and the resource consents or permits system. There is as much an emphasis on monitoring as there is on assessment and planning. These processes are intended to enhance policy formulation and implementation, and make planning more adaptive and responsive to environmental trends and community needs.

The legislation makes environmental policy more comprehensive and integrated than before in four main ways.[19] First, it provides both "substantive co-ordination and co-ordination by common principles" stemming from a single specific purpose; that is, the sustainable management of natural and physical resources. All of the legislative provisions flow from this purpose. Second, it provides a number of procedural means to achieve integration and comprehensiveness (e.g., policy statements, plans, monitoring, consultation, and review). Third, under the old regime, the environmental impact assessment processes and procedures were separated from statutory planning and were project-by-project assessments. Under the *Resource Management Act*, environmental assessment processes and procedures are "integrated with the planning decision making procedures" and must be applied at regional and local levels of government when making policy decisions or issuing resource consents. Finally, the amalgamation of regional, local, and *ad hoc* authorities makes it possible for "policy to be integrated administratively and politically within each district." Within the newly created autonomous regional councils it has enabled a "comprehensive perspective in integrating (resource) policies that transcend" the boundaries of local councils, whereas before a number of ad hoc authorities were each responsible for different resource sectors. But, not all resources are included – energy and mineral resources are omitted.

Sustainable management Sustainable management, the single purpose of the *Resource Management Act*, is defined as "managing the use, development, and protection of natural resources in a way, or at a rate, which enables people and communities to provide for their social, economic, and cultural well-being and for their health and safety while" at the same time achieving three environmental objectives.[20] These are to sustain the potential of natural and physical resources to meet the reasonably foreseeable needs of future generations; to safeguard the life-supporting capacity of air, water, soil, and ecosystems; and to avoid, remedy, or mitigate any adverse effects of activities on the environment. This means that the needs of people and communities must not be fulfilled in a way that makes unattainable the three environmental objectives. Sustainable management is therefore about environmental (ecological and physical) bottom lines and is to be achieved through the assessment of environmental effects. The definition of "environment" is broad and includes reference to ecosystems, people and communities, and natural and physical resources. The meaning of "effects" is comprehensive in scope, including those that are: positive or adverse; temporary or permanent; past, present or future; cumulative; highly probable; and rare, but of high potential impact.

To underscore the seriousness of this intent, it is germane to note that similar principles are in the *Environment Act 1986* and *Conservation Act 1987*. The various procedures and instruments for doing so are imbedded in these three Acts. However, as the Minister for the Environment emphasized:

> The direction and ethic of the Act is clearly encapsulated in section 5(1). It is about promoting sustainable management. Like so much else in this world, the extent to which we achieve that aim is in the hands of human beings who have to live, not just with the constraints of the physical environment, but also with the rules they are prepared to impose on themselves . . . It is to be hoped that planners, lawyers and judges alike will be able to marry the ethic of sustainable management with the practice of living.[21]

Given the importance of this concept and the difficulty of operationalizing it, as with any aspects of sustainability, this aspect of the *Resource Management Act* has engendered, and will likely continue to engender, debate over terminology.

While the *Resource Management Act* is concerned with the promotion of sustainable management of natural and physical resources, the combined effect of this legislation and other government policies forms the basis for a national agenda for sustainable development, involving consideration of broader social and economic goals along with environmental bottom-lines.[22] The policies and plans that regional and district councils are to produce are in principle under-taken as part of other planning processes by these governments. For example,

Plate 3.1 Sustainable management involving open space zoning of a floodplain in Palmerston North, New Zealand (photo by Neil Ericksen)

local councils are required to produce annual plans and may produce corporate or strategic plans, and tribal groups and regional conservancies produce various management plans. These are to be taken into account when regional and local councils prepare policy statements or plans, as called for under the *Resource Management Act.*

Assessing environmental effects The requirement to assess environmental effects applies to both government agencies, when developing policies and plans, and all developers when applying for resource consents or permits covering development activities, land subdivision, water use, discharges into the environment, and coastal activities. Environmental assessment is, therefore, provided for generically rather than in a specified part of the legislation. This signals a major shift in planning philosophy and procedure away from old style zoning and the control of activities under the *Town and Country Planning Act* and *Local Government Act* and onto assessing the environmental effects of activities.

There is also a shift from the presumption that land-use activities are prohibited unless specified, to a presumption that activities are allowed (i.e., no consent required) unless otherwise specified in the district or regional plan. However, to confuse matters, the reverse applies for water and coasts. If a plan does not make any specification, then a water or coastal permit is required.

This curious contradiction stems from all rights to water in New Zealand being historically vested in the Crown (central government) as common property, while land could be privately owned. For contaminant discharges from industrial and trade premises to air or land, permission is required. For discharges from other sources, the reverse applies – they are uncontrolled unless regulated through a rule in a plan.

The change in focus in the *Resource Management Act* towards the integrated planning of natural resources (rather than the built environment) and on controlling the environmental effects of activities (rather than on controlling the activities themselves) requires professional planners and others to think differently. It also necessitates acquisition of new skills with which to carry out the new mandate. Although planning was an important element under the old catchment/water boards regime, the boards were dominated by engineers and slow to employ professional planners and to give them positions of responsibility.[23] Planners, however, fared better under the old united councils, but their numbers were small due to their weak nature, ensured through local council control. With the boards and councils being amalgamated into the new regional councils, the *Resource Management Act* further enhanced the role of planning.

Hierarchical system The *Resource Management Act* provides a framework for integrating national, regional, and local level decisions about resource use, with the main emphasis on empowering regional and local councils to mitigate undesirable human effects on the environment. This entails a variety of entities at the central-government level, and creation of an intergovernmental system for resource management. At the central government level, decisions in the national interest are made in practice by the Executive Council of central government. These include approving national policies and standards prepared by the Ministry for the Environment and the compulsory national coastal policy prepared by the Department of Conservation. The latter approves the coastal plans that are required of regional councils, while the Ministry for the Environment reviews the regional policy statements of regional councils as well as the district plans that are required of local councils. The Planning Tribunal hears complaints and appeals, issues enforcement orders, and penalizes offenders. The Parliamentary Commissioner for the Environment, who is directly responsible to Parliament, has power to review the performance of agencies regarding environmental outcomes under the *Environment Act 1986*, as can the Minister for the Environment under Section 24 of the *Resource Management Act*. The Audit Office may also extend its review of agency performance beyond financial matters.

The legislation establishes a hierarchical system of policy statements and plans as a basis for planning and managing resources among central, regional, and local governments. Central government may issue optional national policy statements and regulatory standards, except for the legislatively required coastal policy statement for which the Minister for Conservation has responsibility. If optional national policies or standards are issued, they are compulsory in guiding the development of resource management instruments lower in the hierarchy.

At the regional level, councils are required to prepare a regional policy statement that provides an overview of resource management issues and policies aimed at achieving integrated management of natural and physical resources in the region. These policy statements are therefore similar to regional plans in Florida. Following from the regional policy statement, regional councils can prepare regulatory regional plans on a range of matters, but these are optional (except for the coastal marine area). Because they provide guidance to constituent local councils, the regional policies and plans are to be developed in consultation with the affected local councils.

Each local council (district and city) is required to prepare a district plan, including rules for guiding development, in order to achieve the purpose and principles of the legislation, especially with regard to managing the effects of land-use activities. These regional policy statements and plans and district plans replace the regional and district scheme plans provided for under the previous planning regime. District plans have some similarity to land-use management regulations and zoning ordinances in the United States. An important difference, however, is that district plans now focus on the effects of activities and their management, rather than the activities themselves. In other words, activities-based regulations are discouraged as being too prescriptive. It is worth emphasizing that plans are not mandatory except in relation to land use, through district plans, and the coast, through the regional coastal plan.

District plans must not be inconsistent with regional policy statements and plans, which must not be inconsistent with national policy statements and standards. Policies and plans have to be publicly "notified," and are open to public submissions and appeals that include comments by central government agencies. In a recent declaratory decision, the Planning Tribunal, which hears appeals and helps establish planning case law, made it clear that hierarchy within the legislation is not to be taken as meaning that local councils are inferior or subordinate in any way to regional councils. Rather, they operate in parallel and in partnership.[24]

In preparing policies and plans, or when making changes to them, central, regional, and local agencies must formally consider a range of policy alternatives and justify them in terms of their benefits, costs, and efficiencies for achieving environmental outcomes. This is called a Section 32 Analysis. In addition, applications for resource consents, such as for subdividing land or discharging into water, must be evaluated in the context of the district or regional plan and sustainable management. Critical to the overall process of management is the requirement that councils implement monitoring systems. These are to assess compliance with conditions on resource consents, outcomes of policies, the state of the environment; and baseline information.

Functions of local authorities In addition to policy and plan preparation, various functions for regional and local councils are specified in the *Resource Management Act*. It is intended that regional councils will: control the use of land for such purposes as soil conservation, water quality and quantity; avoid or mitigate natural hazards; prevent or mitigate the adverse effects of hazardous substances; control the coastal marine area (in conjunction with the Minister of Conservation) for a range of effects, including natural hazards; control pollution onto land or into air and water; and, in relation to any bed of a water body, to control of plantings for various purposes, such as soil conservation, water quality, and avoidance or mitigation of natural hazards. For their part, it is intended that local councils will: control for land subdivision and noise, and any actual or potential effects of the use, development, or protection of land, including for the purpose of avoiding or mitigating natural hazards and the prevention and mitigation of the adverse effects of hazardous substances.

Clearly, some of these functions are discrete to each council (e.g., noise control under local councils and water pollution control under regional councils) and others are overlapping. For example, both councils have responsibility for natural hazards and hazardous substances. In addition, the distinction between "controlling the use of land" for various purposes by regional councils and "controlling for the effects of land use" by local councils is a subtle one since consideration of effects surely also involves consideration of causes or activities. These overlapping responsibilities are a source of potential problems, which we address in Chapter 6. Recognizing this potential, the legislation gives regional councils power to identify which aspects of overlapping functions shall remain within their purview and which shall be the responsibility of constituent local councils, and also grants both regional and local councils power to transfer and/or delegate some functions or to combine to prepare plans.

Implementation design

The architects of the new resource management regime hoped that the freedom given to regional and local councils to prepare their own futures would encourage them to develop sound environmental planning systems and practices. When introducing the reforms, the Minister for the Environment explained somewhat equivocally that "central government will continue to set national standards where these are required. We do not want slack environmental standards to be used as an incentive to attract business development (as) happened overseas . . . " However, he said that central government "does not want to play 'big brother' and prevent people from getting on with the job."[25] Four years after the legislation was enacted, apart from the required coastal policy statement, only one other statement/standard is being prepared. The Ministry for the Environment has produced guidelines (numbering eight by April 1995, with five others in process at that time) relating to various aspects of environmental assessment and implementation of the *Resource Management Act*. These guidelines contain standards, but they are non-statutory in the formal sense; some could eventually be recast as national policy statements.

As required by the *Resource Management Act*, local governments have to first identify issues and problems in their areas and then decide who would be responsible for them. They then have to think through the kinds of environmental outcomes to aim for and how these could best be achieved, including ways for evaluating the means by which results were to be obtained. While regulations could be used in regional and district plans, these were to be only one in a mix of measures that could include public information and education programs, economic instruments, technologies, and environmental standards that could be achieved by any appropriate means.

The choices for meeting the environmental outcomes required of the legislation are left to the respective local councils to make. Policy-makers were apparently willing to tolerate variation among councils in outcomes. They appear to have recognized that poor decisions will result in poor outcomes, which if bad enough will result in either learning to do better next time round, or intervention by the Minister for the Environment. Consultation, cooperation, and sharing between regional councils and their constituent local councils on matters of mutual concern are required for integrated and sustainable management of natural and physical resources to succeed. The participatory features of the planning process are intended to ensure that poor outcomes are not an inevitable result. How this is to be achieved has been left largely to the councils themselves to work through.

The logic is that once councils have established policies and standards for sustainably managing the environment, the consent process would be more straightforward and streamlined for those who need to apply for resource consents or permits for development approvals. These are required under the legislation by a resource developer (the applicant) for uses of land and the beds of lakes and rivers (land-use consent), subdivision of land (subdivision consent), uses of water (water permit), use of the coastal marine area – including discharges and water uses (coastal permit), and discharges of contaminants into the environment (discharge permit). The consent procedure also applies where plans provide that an activity will not be allowed without a consent.

Regional, coastal, and district plans can include rules which prohibit, regulate or allow activities. The rules can group activities into six types

1 permitted – activities are allowed as of right as long as conditions in a plan are met;
2 controlled – application for a consent does not have to be publicly notified, because the activities have minor effects;
3 discretionary – activities are allowed if a consent is given, based on criteria set in a regional or district plan;
4 non-complying – activities that contravene a plan, but are not prohibited;
5 prohibited – activities which a plan expressly prohibits, and no consent can be given; and

6 restricted coastal activities – requiring the consent of the Minister of Conservation, and specified in the regional coastal plan as discretionary or non-complying.

Applications for resource consents that require public notification entail public submissions in writing from anyone on any matter in support or opposition of the application, possible hearings (if the consent authority so deems or if a respondent requests to be heard), and a potential appeal to the Planning Tribunal (if anyone is dissatisfied with a decision) resulting in a rehearing of all aspects of the matter. All of this takes place within deadlines specified by the legislation.

Reforming local government and resource management law suggests that a great deal had to be created in order to implement the new environmental mandate of central government. To a large extent, the rebuilding that has taken place consists of retrofitting existing institutions, laws, and processes to the new requirements. This has made the task of implementing the new policies less daunting than would otherwise be the case. Nevertheless, the fundamental shift from activities-based to effects-based planning, while at the same time adapting to new institutional arrangements and modified processes and requirements, has proved to be a more demanding and costly exercise for regional and local government than was originally anticipated. We examine this transition in Part II of the book.

NATURAL HAZARDS UNDER THE NEW REGIME

The reforms of the 1980s changed the emphasis in dealing with natural hazards. The traditional approach was that of protection against extreme events and provisions for recovery from disasters. The newer approach treats natural hazards as elements of environmental planning and development management within local councils. The following sections describe how this is accomplished. The effects of these changes on environmental sustainability are examined in Chapter 8.

Natural hazards and the Resource Management Act

Natural hazards are not treated separately in the *Resource Management Act*. Rather, they are dealt with through the functions and duties required of relevant agencies and the mechanisms specified for carrying out the duties. As already described, the legislation prescribes a process to be followed when dealing with problems of environmental planning and development control. To this extent, it is not too dissimilar to the predecessor planning and local government acts. Indeed, a number of clauses pertaining to land use and sub-division controls for reducing natural hazards are based on these earlier acts. There are, of course, important differences, but these pertain more to the inter-organizational structures and processes than to substantive prescriptions.

Under the *Resource Management Act*, the Minister for the Environment has authority to prepare a national policy statement for natural hazards, but none is expected. The Minister has set policy that diverts funds away from structural controls and into helping councils to prepare regional policy statements and district plans specific to natural hazards. In addition, grants have been used to encourage exploration of alternatives to regulatory controls, such as economic instruments, for allocating resources. More generally, the Ministry has enjoined other central government ministries in developing policy for disaster and emergency response and in preparing amendments to the *Resource Management Act* on natural hazard issues.

Several policies in the coastal policy statement, issued by the Minister of Conservation in 1994, are directly relevant to natural hazards. For example, the policy states that "local authority statements and plans should identify areas in the coastal environment where natural hazards exist [and] New subdivision, use and development should be so located and designed that the need for hazard protection works is avoided." Where existing use is threatened by a coastal hazard, structural means for protection are recommended only where they are the best practicable option for the future. If they are not practicable, the policy recommends that abandonment or relocation of existing structures be considered. If coastal hazards are unavoidable, the policy states that any

Plate 3.2 Newer housing in Bay of Plenty, New Zealand, at risk from coastal storms and erosion (courtesy of Richard Warwick, Centre for Environmental and Resource Studies, University of Waikato)

actions to address them should avoid adverse environmental effects to the extent practicable.

Requirements for dealing with natural hazards in the *Resource Management Act* are most explicit with respect to procedural requirements for regional and local council decision-making in addressing hazards. This is in keeping with the spirit of government reforms to have local communities live with the consequences of their resource development decisions. There is, therefore, a series of requirements for councils to deal with natural hazards through regional policies, district plans, monitoring, record keeping, provision of information to affected parties, resource consents, subdivision consents, and reserves along the margins of coasts and rivers. Each of these elements in the legislation makes specific reference to avoiding, remedying, and/or mitigating natural hazards.

The legislation does not prescribe how these various requirements for managing development in hazard-prone areas are to be met. Instead, the intent is to allow for innovative approaches and measures, as long as these support the central goal of the sustainable management of natural and physical resources. Because environmental planning under the legislation is outcomes driven, the final test is that in providing for the health, safety, and well-being of people and communities, development does not adversely affect the natural environment.

Natural hazards and the Building Act

The *Building Act 1991* deals with buildings on hazard-prone land in two ways.[26] First, where construction of a building is likely to increase the hazard risk for the land, or any other property, the local council must refuse a building consent unless it is satisfied that there is adequate provision for protection or restoration of property. Second, where the local council is satisfied that the construction will not add to the risk, it may grant a building consent, but the fact that land is subject to natural hazard must be entered on the title of the land. Having done so, the local council is exempt from liability should the building be subsequently damaged by a natural event. In effect, an owner can obtain a building consent to build on hazard-prone land if the building does not add to the risk to that land or to other property, as long as the owner takes the risk and the fact is recorded for the benefit of any future owner of the property.

While the *Building Act* affects the building's construction and subsequent use, the *Resource Management Act* affects the placement of the building as it relates to hazardous land. However, provisions of the former state that no one should be required to achieve performance criteria which exceed the building code – unless other legislation specifically provides for it. This was interpreted as meaning that rules in regional and district plans cannot be more restrictive on buildings than the building-code provisions. This presented policy inconsistency, for example from the fact that the criteria in the building code for surface flooding is the 50-year storm, but many flood-control projects and accompanying building and land elevation requirements, imposed under earlier legislation, are above that level; typically the 100-year event. An amendment to the *Building Act* sought to rectify the anomaly, by allowing local councils to develop rules relating to surface water flooding that are more stringent than the provisions of the building code, but only where the building affects other property, rather than specific to the building itself.[27]

The consequence of having differing objectives under the two acts could be this: some councils emphasize use of the *Resource Management Act* to map and disclose hazard-prone areas so that development can be controlled within them, while some other councils emphasize the use of the *Building Act* to justify building in flood-prone areas by tagging their certificate of titles and letting owners take responsibility for the consequences. The longer-term outcome may well result in a significant variation in hazard-mitigation approaches.

PAST AND PRESENT POLICIES COMPARED

To what extent is the new regime more cooperative and flexible than the old? The most direct comparison that can be made on building and land-use controls is the provisions of the *Resource Management Act* (as supplemented with those of the *Building Act*) as compared with the provisions of the earlier *Town and Country Planning Act* (as supplemented with those of the Local Government Act).[28] Table 3.1 provides a summary comparison of the character of environmental management under the two sets of regimes.

Procedural specifications are extensive under both sets of regimes, but they differ considerably in foci. These differences are reflected in the differences in

Table 3.1 New Zealand's environmental management regimes

Components	Town and Country Planning Acts (circa 1974–1991)	Resource Management Act (1991–present)
Intergovernmental policy	Prescriptive	Cooperative
Sustainability concept	Aspects of sustainable development	Sustainable management of resources
Policy prescriptions		
Procedural	Extensive	Extensive
Substantive	Restrictive and extensive	Flexible and extensive
Sanctions (*vis-à-vis* local governments)	Weak	Weak
Incentives		
Grants and subsidies	Moderate	Limited
Technical assistance	High	Moderate

substantive policy prescriptions relating to the contents of local policies or plans, where the earlier act emphasized activities-based regulations (i.e., prescription). As discussed in this chapter, the provisions of the *Resource Management Act* reflect the contemporary concern for a more flexible approach to planning. Nonetheless, the substantive requirements of the newer are far more extensive than those of the older regime. As such, the new regime is more integrative than the past planning regime.

Neither sanctions nor incentives for local government planning are particularly strong under the *Resource Management Act*.[29] There is review of local policies and plans through the submission process, and central government provides local governments with technical assistance and planning grants. The old regime contained a greater range of incentives for local government compliance than for the new regime. Again, there is a difference in emphasis. The incentives under the old regime tended to be focused on funding actions with a strong focus on building structural works for natural hazard protection. In contrast, the new regime emphasizes technical assistance and funding, albeit limited, for policy and plan preparation. Local councils are left to foot the bill for specific actions in implementing their policies.

The differences in sustainability concepts that are embodied in the two regimes underscore the philosophical differences that are part of the resource management reform. The earlier planning regime focused on "wise use of resources" and attempted to draw attention to a range of environmental and social concerns, recognizing that the choices of any given council were prescribed by both tradition and planning law. This embodied aspects of what would now be called sustainable development. In contrast, the current emphasis is sustainable management of resources and attainment of environmental objectives.

CHALLENGES POSED BY THE NEW REGIME

The *Resource Management Act* is a far-reaching experiment that provides a level of integration in the management of physical and natural resources through the assessment of environmental effects that makes New Zealand a world leader. There are limitations. The legislation excludes aspects of mineral development, energy, and explicit treatment of the traditional planning concerns of housing, transportation and employment. Given these exclusions, it is clear that the legislation reflects a dominance of protection of resources and the environment over those of energy and urban development. Nevertheless, the integrated nature of the legislation is underscored by the fact that it provides a framework both for allocating many natural resources and for managing the adverse effects of the use, development, and protection of land and natural resources. In terms of policy focus, the *Resource Management Act* is not that different from the foci of the previous planning and natural resource management regimes. What is different, however, is the changes in the way business is done as the result of specific provisions of the new environmental management regime and changes in the structure of government. These changes, in turn, raise various administrative and implementation challenges which are examined in Part II of the book.

Administrative and implementation challenges

The newly constituted local governments have a widened range of responsibilities compared with the old, which include for example assessing environmental effects, monitoring policy effectiveness, and enhanced public participation requirements. The history of integrated water management in New Zealand supports the supposition that unless financial resources expand to meet new demands, many local governments will be hard pressed to meet their statutory responsibilities.[30] Lacking substantial new funding, the capacity of councils to adequately execute the new agenda is limited. A reasonable expectation is that there will be an uneven spatial distribution of achievement under the planning regime throughout the country with well-financed local councils doing better than those with fewer resources.

The *Resource Management Act* establishes a need for intergovernmental co-ordination and cooperation through its three-tiered hierarchical structure. The philosophy behind the creation of the regional and local councils is one of intergovernmental cooperation in carrying out complementary functions. The framers of the legislation were clear that regional councils were not to be viewed as being superior to local councils. However, the legislation is unclear as to how this dialog will take place except to say, for example, that district plans are not to be inconsistent with regional policy statements. Effective policy coordination requires elected and profession staff of both sets of councils to meet regularly and to agree on processes for achieving mutually beneficial outcomes. Given the demanding statutory deadlines for mandated policies and plans, not to mention traditional animosities between levels of government, it is possible that such harmony will not universally result. Far from breaking down the old dichotomies in institutional cultures (regional–local and urban–rural), it is possible that the schisms will grow under the new regime.

The efficacy of the institutional arrangements and procedures of the *Resource Management Act* will be largely dependent on the participants in the planning process. Of critical importance will be the commitment of local government officials to the goal of sustainable management, and their ability to avoid the parochialism of past practices. Also relevant is the extent to which policy statements and plans reflect a change of thinking on the part of councillors, planners, and other professionals involved in the process in moving from activities-based prescriptions to outcomes-based assessments. These pose noteworthy challenges since the legislation calls for the application of new tools and methods to the assessment of environmental effects that are as yet weakly developed. The danger is that practitioners will continue to muddle through in the absence of well-developed methods and a willingness to learn to use them.

The *Resource Management Act* is at a critical stage. It will be five to ten years before a new system of operative regional policy statements and regional and district plans is in place. One question is whether central government politicians are prepared to wait that long for results. If they are not, there is a prospect that the legislation may change. Change may, however, come as a consequence of the newly elected system of mixed membership proportional representation of Parliament. In addition, a hands-on Minister for the Environment could contribute to significant policy development. Ultimately, how the legislation plays out will depend on how various players respond to the challenges and opportunities that the new regime poses.

Sustaining sustainability

New Zealand has always been quick to adapt to relevant world trends and on occasions has led the way. When conditions required, it has demonstrated a remarkable ability to radically change course. However, the concept and practice of sustainable management of resources evolved over a long period in New Zealand and has included such experiments as nature preservation through development of a national park system from 1886, dynamic conservation of forest resources from the 1920s, comprehensive catchment treatment and control from the 1940s, rural and urban development controls from the 1950s, and environmental protection and enhancement from the 1960s. Each of these approaches attempted to buffer the adverse effects of the private and public development of the nation's natural and/or physical resources, while at the same time improving public welfare. But the results were mixed at best, and at times dismal.

What is different about the 1990s is that the sustainable management of natural and physical resources has been integrated into reforms of local government and resource management law. These, in turn, form part of the wider administrative, bureaucratic, economic and social reforms of the 1980s underpinned by New Right ideology. In this new regime, resource developers are free to exploit resources for their self-interest, so long as they internalize the environmental costs of their production and meet the environmental bottom-lines that are being established by regional and local councils. For their part, the councils are encouraged to approach resource management in a flexible and innovative manner with minimal interference (and some would say, help) from central government. In return for this freedom, it is assumed that councils and developers will act in an environmentally responsible way. It is also assumed that the new effects-based environmental assessments and associated monitoring at local level will in time combine to enhance the overall quality of the New Zealand environment.

This new approach to the sustainable management of resources means that participants in the process are on a steep learning curve. It remains to be seen whether this bold experiment achieves its broad goals. Or, whether problems like rabid parochialism, lack of resources, and limited direction from the central government cause yawning gaps to appear in the sustainable management of resources across the nation.

NOTES

1 Professor John Roberts quoted in Roger Blakeley, "Water in Society: Policy and Practice," *Water Conference 1988: Papers of the Fifth National Water Conference*, sponsored by the Institution of Professional Engineers New Zealand and the Royal Society of New Zealand, Dunedin, New Zealand, August 15–19, 1988 (Wellington: Institution of Professional Engineers, 1988), pp. 3–28, at 3.

2 Provincial government was disestablished when local government was established in 1876 by the *Municipal Corporations Act and Counties Act*. The territorial local authorities were given power to collect a land tax (rates) from property owners to carry out community services.

3 See, for example, Roger Wilson, *From Manapouri to Aromoana: The Battle for New Zealand's Environment* (Dunedin: Earthworks Press, 1982).

4 An example of the complexity with regard to water and land management is given in Neil Ericksen, "Water Management in New Zealand: Evolution or Revolution?" in *Integrated Water Management: International Experience*, ed. Bruce Mitchell (New York: Belhaven Press, 1990), pp. 45–87. For coastal management see: Jack W. Lello, "Seminars: Coastal Management in the 80's," *New Zealand Environment* 28 (Summer 1980): 6–11.

5 These provisions were part of the *Town Planning Act 1926*. Prior to 1989, territorial local authorities included counties, boroughs, towns, and cities. Since 1989, various local authorities have been amalgamated into what we label for simplicity as local councils. These include district councils and city councils. Local councils can be distinguished from the newly constituted regional councils.

6 Bill Williams, *District Planning in New Zealand* (Auckland: New Zealand Planning Institute, 1985).

7 *Town and Country Planning Regulations 1960* quoted from Neil J. Ericksen, *Creating Flood Disasters?* Miscellaneous Publication No. 77 (Wellington: National Water and Soil Conservation Authority, Water and Soil Directorate, 1986), p. 148.

8 A detailed account of urban floodplain management under the pre-1990s' regime is given in Neil J. Ericksen, *Creating Flood Disasters?*, op. cit.

9 See Roger Douglas, *There's Got to Be a Better Way: A Practical ABC to Solving New Zealand's Major Problems* (Wellington: Fourth Estate Books, 1980). Roger Douglas was a prominent member of the Labour Party who fell from grace in the late 1970s over policy, and wrote the aforementioned book. He regained favor, becoming Minister of Finance in the Fourth Labour Government, 1984–7. He was the driving force behind the reform movement (Rogernomics), which ran counter to the traditional socialist values of the Labour Party. He lost power in late 1987 and eventually set up his own party (Association of Citizens and Taxpayers – ACT) to pursue his vision of a tax-free New Zealand.

10 A poorly performing economy was made worse when the outgoing Prime Minister and Minister of Finance, Robert D. Muldoon, refused to devalue the New Zealand dollar after losing a snap election in 1984. It caused a dramatic and near calamitous run. Central government intervention culminated in a wage–price freeze in 1982 aimed at controlling inflation. The environmentally harmful projects of central government for the period 1968 to 1982 are cataloged by Roger Wilson, *From Manapouri to Aramoana: The Battle for New Zealand's Environment*, op. cit.

11 For detailed accounts see Ton Bührs and Robert Bartlett, *Environmental Policy in New Zealand: The Politics of Clean and Green?* (Auckland: Oxford University Press, 1993), and Ali Memon, *Keeping New Zealand Green: Recent Environmental Reforms* (Dunedin: The University of Otago Press, 1993).

12 The Ministry for the Environment and Parliamentary Commissioner for the Environment were created by the *Environment Act 1986*, and the Department of Conservation was created by the *Conservation Act 1987*.

13 This was achieved through the *Local Government Amendment Act [No. 3] 1988* and *Local Government Amendment Act [No. 2] 1989*.

14 Ton Bührs and Robert Bartlett, *Environmental Policy in New Zealand: The Politics of Clean and Green?*, op. cit., p. 121.

15 For a useful review of past and present planning regimes see P. Ali Memon and Brendon J. Gleeson, "Towards a New Planning Paradigm? Reflections on New Zealand's Resource Management Act," *Environment and Planning B: Planning and Design* 22, no. 1 (1995): 109–124.

16 Rt Hon Geoffrey Palmer [Deputy Prime Minister and Minister for the Environment], "Resource Management Law Reform – A Progress Report," speech given to the New Zealand Catchment Authorities Association Conference, Hamilton, New Zealand, 19 April 1989.

17 United Nations, Organization for Economic and Cultural Development, *Environmental Policies in New Zealand: A Review by the OECD and its Environment Committee* (Paris: Organization for Economic and Cultural Development, 1981).

18 Maori are the indigenous people of New Zealand with whom a treaty was struck by the British Crown in 1840. They make up about 15 percent of the total population of 3.5 million people.

19 The commentary in this paragraph draws from points made by Ton Bührs and Robert Bartlett, *Environmental Policy in New Zealand: The Politics of Clean and Green?*, op. cit., p. 147.

20 Quoted from Section 5 of the Act. The interpretation of the meaning of sustainable management in the text is based upon Arnold R. Turner, "Memorandum for the Representatives of the Minister of Conservation, and All Other Parties Interested in the Matter of the *Resource Management Act 1991* and the Inquiry into the Proposed New Zealand Coastal Policy Statement" (Wellington: Board of Inquiry, 1993). But see also John R. Milligan, "Pondering the 'While'," *Terra Nova* 16 (May 1992): 50–51.

21 Honorable Simon Upton [Minister for the Environment], "The Resource Management Act, Section 5: Sustainable Management of Natural and Physical Resources," keynote speech to the Second Annual Conference of the Resource Management Law Association, 7 October 1994 (Wellington: Ministry for the Environment, 1994), p. 10.

22 See: Ministry for the Environment, "The Context: Sustainable Development as a Backdrop for the Resource Management Act, Information Sheet Number One" (Wellington: Ministry for the Environment, 1992). The meaning of sustainability in the *Resource Management Act* is subject to debate; see for example: Janet McLean, "New Zealand's Resource Management Act 1991: Process with Purpose?" *Otago Law Review* 7, no. 4 (1992): 538–555; and Bruce V. Harris, "Sustainable Management as an Express Purpose of Environmental Legislation: The New Zealand Attempt," *Otago Law Review* 8, no. 1 (1993): 51–76.

23 This professional boundary problem in water and soil management is described in Neil J. Ericksen, "New Zealand Water Planning and Management: Evolution or Revolution?" op. cit.

24 Decision No. A89/94 in the matter of the *Resource Management Act 1991* and an

application by the Canterbury Regional Council under Section 311 for declarations, Planning Judge D.F.G. Sheppard presiding (Christchurch: The Planning Tribunal, 2–3 November 1994), pp. 7–8.

25 Rt Honorable Geoffrey Palmer (Deputy Prime Minister and Minister for the Environment), "Resource Management Law Reform," Speech given to the New Zealand Catchment Authorities Association Conference, Hamilton, 19 April 1989 (Wellington: Ministry for the Environment, 1989).

26 The *Building Act 1991* provides a new framework for the control, construction, and maintenance of buildings. The Act and subsequent building regulations adopted in 1992 establish a national building code which supersedes that of all previous building regulations and bylaws of local authorities. The building code contains mandatory performance-based minimum requirements that specify how a building must perform to ensure that the health, safety, and amenity needs of people are met, along with requirements that other property is protected from damage.

27 The relationship was subject to further review, for which a declaratory judgement of the Planning Tribunal enhanced rule-making under the *Resource Management Act*. However, this declaration was still subject to appeal at the time of our research.

28 Note that in addition to the planning acts, the water and soil legislation was especially important for managing flooding and erosion under the old planning regime. Aspects of these are still relevant under the new regime, but have lesser importance because central government subsidies for new structural works have been eliminated. Civil defense legislation was, and still is, relevant for community preparedness and emergency response to natural disasters.

29 While incentives and sanctions for councils are not particularly strong under the *Resource Management Act*, penalties for non-complying resource users are very strong in comparison to the previous planning regime. For example managers and directors are liable for the acts of their agents and imprisonment is the toughest penalty.

30 For elaboration, see Neil Ericksen, "New Zealand Water Planning and Management: Evolution or Revolution?" op. cit.

4

TOWARD COOPERATIVE POLICIES
Flood management in New South Wales

The history and culture of non-Aboriginal Australia is intimately entwined with the struggle against a harsh environment. Indeed, there is a national pride in the fight against environmental hazards symbolized by the image of the "Aussie battler" – a farmer struggling against flood or drought. There is the reality of a harsh Australian natural environment behind this mythology. Severe flooding was reported from the earliest days of the colony, with the governor of the early 1800s, Lachlan Macquarie, attempting to direct settlement away from flood-prone areas. Despite his enormous executive power, Macquarie was unable to influence private development, much less control it. Within a few years, following further floods, he had reason to chastise settlers for "their wilful and wayward habit of placing their Residences and Stock Yards within the reach of the floods" and for their failure "to remain within the flood marks of the townships assigned for them on the high lands."[1]

More than a century and a half later, policy-makers in Australia were trying to move beyond mere acknowledgment of serious problems in the natural and built environments. By the 1970s, recognition of environmental harms and the influence of trends overseas, particularly in the United States, saw the development of environmental policies in many Australian states that went well beyond exhortation. The states have direct authority over most land-use and natural resource issues. However, local governments act as the agents of state government on many matters, including control of land use. Planning powers are local but because of state authority to intervene, these powers are shared with the state governments.

As a consequence, a central challenge for Australian states is to find ways of getting environmental objectives met by local governments. Some efforts have been strongly prescriptive, including the precursor of the flood policy of the state of New South Wales that we examine. But given a tradition of largely cooperative approaches to intergovernmental policy, while reserving potential for state or commonwealth intervention, highly prescriptive and coercive approaches are uncomfortable for Australian policy-makers. The New South Wales flood policy serves as an example of a more flexible intergovernmental approach to environmental management.

Plate 4.1 Urban flash flooding in Australia (courtesy of the Disaster Awareness Program, Emergency Management Australia)

The broader issues of environmental management are intertwined with efforts to develop sensible policies for floodplain management. As such, flood policy, particularly as it has evolved in New South Wales, serves as a microcosm of larger conflicts. There are tensions between levels of government and between conflicting objectives within levels of government. With respect to sustainable development issues, efforts to manage flood hazards highlight conflicts between land-use controls for protecting areas and the contradictory aim of promoting development. In addition, there are historic tensions between local planning responsibility, state flood policy, and funding arrangements and technical expertise favoring engineering approaches to flood problems.

This chapter examines the evolution of flood policy in New South Wales, beginning with the efforts to establish land-use policies in the early days of settlement, followed by pressures for development, later efforts to restrict development, and the current flood policy. In documenting these changes, we address broader issues of governance and planning processes as they relate to environmental management more generally. The chapter concludes with the challenges likely to be faced by those charged with policy design and implementation, especially in the context of the imperative for sustainable development.

70

THE SETTING

In terms of both policy development and extent of flood hazard, New South Wales is a logical choice for study. Figure 4.1 places New South Wales within the Australian setting. Along with the state of Queensland, it shares the nation's worst flood problems. A national review of flooding undertaken in 1992 showed substantial flood potential in New South Wales, exceeded only by that of the state of Queensland.[2] Sydney, the state's capital and the nation's largest urban area, and its many metropolitan jurisdictions present a substantial urban flood problem. There are a total of 177 local governments (henceforth, councils) in New South Wales, comprised of 31 municipalities, 35 cities, and 111 shires. By virtue of the extent of policy innovation and the fact that it is the only state with a well-established flood policy, New South Wales is the leader in flood policy development in Australia.

FLOODPLAIN MANAGEMENT: SEEKING A POLITICALLY SUSTAINABLE APPROACH

Flood policy with a strong planning theme has a long history in New South Wales. The first documented flood occurred on the Hawkesbury River in 1799. Lachlan Macquarie's exhortations are generally regarded as the first attempt to

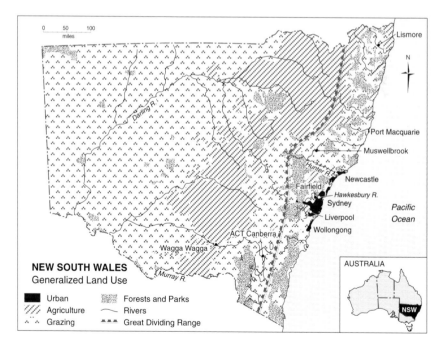

Figure 4.1 The New South Wales setting

71

develop a flood policy for the state. He tried to persuade people by proclamation to build in flood-free areas. In 1810, he founded six townships, all sited with the principal criterion that they be flood free.[3] But his efforts met with limited success in guiding development toward flood-free locations during this critical period of initial settlement and consolidation. Economic and development considerations dominated. Many of the flood-prone towns of today are a legacy of this failure, with the irony that the economic rationale for their location has long gone.

Emergence of local planning

With the start of the twentieth century and federation came broader powers vested in local governments. Their ability to influence land use remained very limited because strong land-use powers were reserved for the states. With the enormous expansion of government during the 1930s and the Second World War, it was logical that local planning powers would be firmly established. Legislation to this effect was passed in 1945 as an amendment to the New South Wales *Local Government Act*. This gave local governments some control over land use, including that in floodplains, through powers to approve subdivision of land, to zoning based on an official plan, and to regulate building.

Although some councils did not complete their zoning schemes until the early 1970s, they were able to manage development through interim development regulations.[4] Many councils passed building ordinances requiring floor levels to be a certain height, but monitoring for compliance and enforcement appears to have been weak. Local officials did not generally desire or act to inhibit development since they wanted to promote economic growth. They saw their planning role as largely a technical function, limited to the provision of the infrastructure necessary to accommodate growth.

State attempts at flood-related planning

Most local governments did not use the various planning provisions to prevent the development of flood-prone areas. After severe flooding in the 1950s raised awareness of flood hazards, local officials pressed for structural solutions that would "protect" citizens from flood hazards. Although some post-flood reports argued for controls over development in flood-prone areas, the reluctance of many local officials to undertake such land-use planning is evident from their unwillingness to undertake any form of land-use zoning.

After the floods of the 1950s, the New South Wales Department of Local Government issued a brief policy statement that referred to the need for floodplain development control.[5] Like Macquarie's efforts, the circular relied on obtaining local government action through persuasion. There was no attempt to build local government commitment or capacity for undertaking the recommended restrictions on development, compounded by the fact that the legal basis for the circular was uncertain. Indeed, the restrictions that the state imposed on planning were contradicted by the actions of county councils. These are single-purpose entities, with funding from constituent local governments and the federal government, that serve as construction authorities for flood protection and other structural works.

The county councils presented a case for federal financial assistance at their 1963 conference. Subsequently, the Commonwealth passed the *New South Wales Flood Mitigation Grants Act* that provided funds initially for six years, and through extensions to the present, on a formula of two-part federal, two-part state, and one-part local funding. Because the federal government did not want to be involved in approving individual projects, the funding for structural works was effectively under state control. The association of flood mitigation authorities that was created in the late 1950s to lobby for funding has subsequently expanded membership, and has remained an effective lobby and networking group.

At the time of creating these funding programs, all moneys passed through the county councils who could overrule their constituent local councils on matters relating to flood mitigation. This reinforced the view that flood management was an engineering matter. For the first decade or so, funds from this source were exclusively for construction. Only a very few councils, mainly located in areas without county councils, implemented flood-related planning measures without state assistance.

Prescriptive planning: the 1977–84 flood policy

Destructive floods of the early 1970s stimulated a reappraisal of the state's approach to reducing flood damages. By this time, most of the structural flood-control works that were recommended in the post-1950s flood reports had been constructed. The first group of these were constructed on the coastal floodplains and exhibited favorable cost–benefit ratios. But as evidenced by the flooding of the early 1970s, the costs or disbenefits seemed to grow for subsequent works. In many areas, urban development was expanding onto floodplains. Requirements for floor levels relating to a historic flood were widespread, but they were rarely enforced. It was abundantly clear that structural solutions were insufficient and land-use planning with respect to floods was rare.

The reappraisal led to the development of a state policy on flood-prone land that was initially issued by the state's Planning and Environment Commission in 1977 and clarified by two later statements.[6] The final circular about the policy, issued in early 1982, establishes a policy that "promotes the removal of urban development from flood prone areas wherever this is practicable and appropriate and aims to clear floodways of unnecessary obstructions to the free flow of flood water."[7] Floodplains were defined by the 100-year flood and floodways by a 20-year flood, unless otherwise determined by the state's Department of Public Works or Water Resources Commission. These provisions were to be applied uniformly statewide, ignoring considerable variability in flood exposure.

The policy was mandatory only for state government authorities. But because local governments have primary responsibility for land-use decisions, local government cooperation was necessary for successful implementation. In an attempt to obtain the requisite cooperation, the state government provided technical assistance for flood risk delineation. For over seventy local government areas, state authorities undertook mapping of the different levels of flood risks. These maps were to form the technical basis of local land-use and building regulations, public information, and other aspects of implementing the state's flood-free land policy.

The policy emphasis was clearly prescriptive, in stating what development was permissible, and coercive in that councils had to adhere to the policy in order to avoid undesirable consequences. One sanction was the potential loss of state subsidies for the development of urban infrastructure, affecting services ranging from roads and mortgages to health centers and swimming pools. The prospect of losing this funding put great pressure on councils to avoid allowing further development in flood-prone areas. Another sanction was potential state pre-emption of local planning power. Of most concern to local officials was the potential legal liability they would incur if they did not follow the policy. Local officials were reminded of this liability potential in one of the follow-up circulars to the original policy which states "[if a council] approves development within a flood prone area which is subsequently damaged by a flood, it may be liable to legal action for damages on the basis it was negligent in giving the approval. This relates to the council's duty of care in exercising any of its powers."[8]

Local government cooperation in enforcing the state's flood-prone land policy, through planning controls, was the key to implementation of the regulatory regime. Given the various sanctions for failing to comply, state officials presumed that local councils would adopt and enforce the requisite development regula-tions. The policy had discretionary elements and caveats. However, the zeal with which some state agency bureaucrats implemented the policy ensured that these were often forgotten and that the policy was seen to have only the singular aim of achieving flood damage reduction through planning controls – regardless of the cost. Many local officials resented the prescriptive nature of what they viewed as an impractical policy.

Local government cooperation was not always forthcoming and state officials understandably had difficulties in deciding how to respond.[9] The state policy was undermined in a variety of ways including some local councils referring every development application to the courts (to avoid liability), by token compliance with flood-risk notification requirements, and by the lobbying of local officials for changes in the policy. Increasing political pressure in some localities and insufficient state or national funding to support implementation became serious barriers to the willingness of local councils to cooperate.

By 1979, the policy came under increasing attack by land and development interests, as local officials sought clarification of the legal issues and, in concert with developers and property owners, expressed concern at the policy's fiscal and land-use implications. The strongest political reaction, however, followed the release of flood maps for Sydney's western suburbs in November 1982 and subsequent notification by letter of the flood-prone status of 3,500 individual property owners. The Fairfield Council wrote to these property owners to protect themselves from possible legal action for failing to advise residents of the flood risk. The letters galvanized public opinion against the policy. Many residents believed that they were seriously disadvantaged by the release of flood maps and associated policy provisions. Although moderately severe flooding occurred in 1956, the majority of residents had no personal flood experience at the time the flood maps were issued.

Negative public reaction grew during the latter part of 1983 to culminate just before the March 1984 state election with intense lobbying by an action group of local residents and solid support for policy abandonment by the state parliamentary opposition. A campaign leaflet that was delivered to houses in a number of Sydney suburbs exploited the public's concern over the possible effects of flood maps on property values. It suggested that "[o]nce the Labor Government has mapped your area – and they could be mapping it now – the value of your property could be reduced by up to 50 %," adding that "If your street is declared flood prone you would not be able to sell your house, you will not be able to extend your home, you will not be able to borrow money on the security of your home, [and] you will not be able to insure your home against flooding."[10] The Liberal Party promised to rectify this situation if elected to government, with a specific promise to abolish the standard of the 100-year flood.

In an attempt to counter this campaign material, the Minister for Water Resources, whose electorate included Fairfield, issued a reassuring letter to residents of Fairfield. But, faced with the possibility of the flood policy becoming a major issue only weeks before the election, state Premier Wran pre-empted the opposition by announcing that he would change the policy. He referred specifically to the abolition of the 100-year flood standard and the cessation of state mapping of flood risks. After winning the election, the Wran Government carried through on these promises and replaced the flood prone land policy with the flood "merits" policy on 11 December 1984.

Broadly speaking, the experience and fate of the flood policy from 1977 to 1984 highlights the difficulties faced in any zealous application of land-use controls in Australia. A strong component was the legal uncertainty that precipitated Fairfield Council's notification of flood risks. Another factor was failure to apply the provisions of the flood policy according to the degree of flood risk, treating all areas within the defined standard as equivalent. As a consequence, policy provisions tended to be applied uniformly to the entire legally defined floodplain rather than differentiating areas of especially high risk. Another problem was that the policy failed to recognize the practical need for special provisions for localities that had business districts located largely within floodways or flood-prone areas. Officials in such localities saw the policy as choking off potential for future economic development.

The 1984 merits policy: toward a cooperative approach

In view of the reasons underlying reactions against the policy that existed from 1977 through 1984, and Premier Wran's re-election promises, it is hardly surprising that the December 1984 flood policy statement established a new flood policy and abandoned flood mapping. The new policy embodied a set of philosophical changes in thinking about environmental management as well as new approaches to managing flood hazards. The new policy presents a stark case of policy reversal that resulted in new ways of doing business, rather than simple reversion to old ways. These new ways – the merits flood policy – are the foci of our study of cooperative intergovernmental environmental management in New South Wales. Table 4.1 compares key characteristics of the pre- and post-1984 flood policies.

In terms of overall philosophy, the adoption of a new policy in December 1984 marks a shift to a predominantly cooperative approach to intergovernmental regulation. The state's emphasis is regulatory goals, rather than prescribed standards, under the presumption that local councils can devise the best means within their communities for reaching such goals. The new policy shifts the main responsibility for ensuring sound floodplain management from the state to local governments. The policy also reduces the emphasis on local government restrictions over land use as a means for reducing flood losses. Within the constraints of rules established by local councils, development proposals are to be treated on their merits with respect to a range of considerations.

In contrast to the policy from 1977 through 1984 that emphasized avoidance of development in floodplains and uniform application of policy, the new policy "recognizes, firstly, that flood liable land is a valuable resource which should not

Table 4.1 New South Wales' flood policies

Components	Flood-prone land policy (1977–84)	Flood "merits" policy (1985–present)
Intergovernmental policy	Prescriptive and coercive	Cooperative
Sustainability concept	Environmental sustainability	Ecologically sustainable development
Policy prescriptions Procedural Substantive	Restrictive and extensive Restrictive and extensive	Flexible Flexible but limited
Sanctions (*vis-à-vis* local governments)	Strong	Weak
Incentives Grants and subsidies Technical assistance	High High (mapping)	High High (flood studies)

be unnecessarily sterilized by preventing development, and secondly, that by carefully considering the circumstances of each proposal, developments can be allowed which might otherwise have been unnecessarily refused or unreasonably restricted."[11] In contrast to the previous policy which emphasized only flood frequency as a basis for standards, the merits policy asks local governments to consider potential flood damage and economic, social, and environmental factors in developing local rules about development and land use.

This reflects a change in policy objectives that reflects different concepts of sustainability. The prior policy had the single objective of protecting flood-prone areas and as such can be characterized as incorporating the goal of environmental sustainability. The merits policy has multiple objectives noted above and incorporates what Australians refer to as "ecologically sustainable development." This Australian version of the concept of sustainable development entails consideration of the benefits of environmental protection and consideration of the economic and social benefits of growth. In incorporating this as a policy objective, the merits approach to floodplain management is more flexible than the prior regime in allowing local governments to make hard choices about growth and floodplain management.

A central challenge posed by the merits policy is how to achieve the necessary local government cooperation. A key feature of the policy design is building the capacity of local governments to achieve the policy goals. The main instrument for this is the advice contained in a state publication, originally titled the *Floodplain Development Manual*. It provides guidance for policy implementation and serves as a key source of technical assistance for the development of floodplain management plans. Additional capacity-building features consist of financial subsidies for carrying out flood studies and developing flood programs.

There are three key incentives for local government to comply with and implement the state's merits flood policy. These are strong support from state agencies, resolution of the legal uncertainty, and a flexible negotiated approach. Two key state agencies, the departments of Public Works and Water Resources, provide financial subsidies and technical advice as they did under the previous policy but availability of funds for implementing flood programs is conditioned on local preparation of a floodplain management plan. The legal situation was clarified by an amendment to the local government legislation that partially waived liability for flood-related land-use decisions by councils. This indemnification requires that councils exercise their "duty of care" in "good faith" by acting substantially in accordance with the policy as set out in the state's floodplain development manual.

Flexibility has been introduced through explicit recognition of the need to negotiate tradeoffs between economic, environmental, social, and flood-hazard factors. More specifically, instead of rigid probability criteria, local variation in definition of flood risks is encouraged based on detailed hydrological and damage studies. The state government no longer restricts development subsidies to flood-free land, thereby lifting various restrictions on financing development within flood-prone communities. Other policy provisions provided for reduction in local taxes where vacant land cannot be developed because of flood hazards, thereby addressing an equity issue which was an important impediment to flood-plain regulation.

Despite the cooperative emphasis of the flood merits policy, state agencies still retain oversight of local government development decisions and planning (functions performed by the state's departments of Planning and Local Government). The state government has power to intervene in the planning process and in extreme cases can appoint an administrator to perform the planning (or other) functions of local councils. Such action is politically unpopular and rare, and as such is probably not an effective deterrent. More commonly in recent years, state ministers have intervened in development decisions that were deemed to be of statewide significance.

FLOOD POLICY AND ENVIRONMENTAL PLANNING

The merits flood policy is best viewed as an overlay to the environmental planning system of New South Wales. Local councils are advised to develop environmental plans and regulations that support floodplain management plans. However, flood-hazard management activity has been undertaken without such integration. For example, levees have been routinely constructed to protect existing development without consideration of the broader environmental planning issues raised by such development. More generally, some officials have highlighted the lack of integration between the responsibilities of local government with respect to flooding and the social policies promulgated by the state's planning and housing departments.[12] These tensions are best understood in the context of the state's environmental planning provisions.

Environmental planning provisions in New South Wales

With the 1970s and the initial surge in Australian environmentalism, attempts were made to broaden planning from the location of buildings and associated infrastructure to a comprehensive planning approach encompassing all development, social equity, and the environment. The primary legislative expression in New South Wales was the *Environmental Planning and Assessment Act 1979*. The legislation's objectives are to promote social and economic welfare and a better environment; to share government responsibility for environmental planning with local governments; and to increase the opportunity for community involvement in environmental planning and assessment.[13] Like the flood merits policy, the planning act seeks intergovernmental cooperation and balanced decision-making. Indeed, some administrators refer to this as a merits approach to planning.

To help achieve its objectives, the planning legislation introduced a three-tiered system of environmental planning instruments for which each instrument has a formal legal status. The top tier is a state environmental planning policy. These deal with matters that, in the opinion of the relevant minister, are of state significance. State planning policies with clear implications for floodplain management include a policy that allows the state's housing commission to approve housing in residential areas without obtaining local development consent. Another policy provides special procedures for processing applications for development which will employ 100 or more people. Rather than setting forth comprehensive policy statements, some argue that the state environmental policies tend to involve the ad hoc concerns of state government.[14]

The second tier of environmental planning instrument is a regional environmental plan. These plans or policy statements are produced by state officials, although in practice the affected local governments may have considerable influence in their formulation. A region is anything that the planning minister decides it is. Some regional environmental plans deal with strategic planning for a geographical region such as the Hunter Valley. Other regional environmental plans address more specific issues, such as the development of Kurnell Peninsula south of Sydney.

Local environmental plans (essentially, local land-use and development-control regulations) constitute the third tier and serve, in principle, as the linkage between local floodplain-management plans and environmental regulations. The local environmental plans are prepared by local governments. Unlike American comprehensive plans, which typically address the needs of a given jurisdiction and set forth planning policy, the local environmental plans in New South Wales address specific topics and/or geographic areas. As a consequence, a given local government may have many local environmental plans – some councils have as many as 200 of them. Councils tend to add new environmental plans as new issues crop up, rather than going through a cumbersome process of amending existing environmental plans. In addition to these plans, councils can achieve regulatory goals through use of development-control plans that serve as guides, rather than binding rules, for negotiating development outcomes. These are entirely under the control of councils and do not have the legal standing of environmental policy plans. Because they are simpler to prepare and can be used for achieving regulatory goals, while also allowing discretion, the development-control plans are commonly employed by local councils.

The environmental planning process in New South Wales is subject to several forms of review and/or intervention. Public participation requirements are extensive and any person can make a submission to the council about a draft local or regional environmental plan. Those preparing plans must consult with the public agencies that might be affected by the plans. The planning minister may amend the plan or defer aspects of it although the courts have limited this power for local plans by ruling that a local environmental plan is a "creature of the council." Disputes about the planning process or decisions based on particular plans are normally heard by the Land and Environment Court. Approximately 2 percent of all development applications come before the Court, and about half of these proceed to a hearing with about half of the appeals dismissed and half upheld.[15]

The *Environmental Planning and Assessment Act* reflects the long-standing tension between decision-making at the local level and centralized control by state government.[16] The trend is to make it easier for the relevant minister to decide planning decisions, as illustrated by amendments in 1985 that significantly increased the powers of ministers *vis-à-vis* local governments and the public. Under these provisions, the state government may approve prohibited development where the development is, in the minister's opinion, of state or regional significance. The alternative was seen as the state being, to quote the planning minister during debate in parliament, "deadlocked by a local government supporting parochial interests." This greatly facilitates state approval of locally contentious development proposals. Not surprisingly, the pro-development interests are strong supporters of these changes and the results of such state action rarely have positive environmental outcomes.

Floodplain management and environmental planning

The pre-merits flood policy of 1977–84 attempted to link development control explicitly with flood hazards for which, as discussed earlier, the policy proved to be politically unsustainable. This highlights a fundamental tension in floodplain management between the desire to promote growth and the need to protect hazard-prone areas. This results in conflict between the interests of the development community and those trying to control development for environmental and community reasons. In New South Wales, each group can draw on different sets of legislation and policy to buttress their case.[17]

There are many strategies available to government authorities attempting to manage development in hazard-prone areas. The merits flood policy recognizes the multiple approaches, but does not provide an integrating mechanism for floodplain management and environmental planning. In principle, the primary integrating mechanism is the actions local governments take under provisions of the *Environmental Planning and Assessment Act*. But a strong theme of New South Wales flood management history is the marginal status of planning. At present, there is weak linkage between the state floodplain policy and planning system. Indeed, guidance issued by the state's Department of Planning, restricting development in flood-prone areas (Guidance Memorandum G25), is highly prescriptive and contradicts key premises of the flood merits policy.

Despite these policy inconsistencies, the environmental planning process in New South Wales has potential for greater integration of flood, environmental, and social concerns. Any major engineering work proposed for flood mitigation would likely require preparation of an environmental impact statement under planning act provisions. Flooding is one of a number of issues that would be addressed as part of the environmental assessment. In addition, the flood merits policy guidance specifies that appropriate land-use and other regulatory instruments should be employed to implement floodplain management plans. But, as noted earlier, few local governments appear to have formally incorporated flood elements into local environmental plans. The preferred route is the use of less binding development control plans.

Other institutional mechanisms in New South Wales have evolved which perform, or potentially could perform, an integrating function for various elements of flood hazard and environmental management. These include state interdepartmental committees that include the state Flood Warning Consultative Committee and the State Emergency Management Committee. The portfolio of the latter committee has been extended to include environmental matters, focusing primarily on toxic and other hazardous spills. Committees are a feature of Australian government and hazard management is no exception. They are seen as ensuring inter-agency contact, as a way of making the system work.

IMPLEMENTING THE MERITS FLOOD POLICY

Policy implementation of the state's flood merits policy is guided by the state's *Floodplain Development Manual.*[18] The manual provides detailed guidance for local government preparation of floodplain management plans, specified as a series of steps to be followed in plan preparation. The first step is the establishment of a floodplain management committee that is to provide a forum for policy development and guidance. The guidance does not prescribe a fixed membership for the committee but envisions it as involving representatives of various stakeholders, including general public representation, along with participation by representatives of relevant state agencies. (In some localities, these committees were operating before the new policy was announced.)

The technical basis for flood policy is completion of a flood study comprised of detailed hydrological analyses and consideration of the potential for extreme floods. Detailed information from the flood study on the nature and extent of the local flood hazard is to form the basis for selection of a flood standard for the community or a set of standards for different parts of the community. The pre-merits flood policy specified a uniform standard (the 100-year flood) whereas the merits policy provides councils with flexibility to select a variety of standards according to the nature of the flood hazard and community values concerning social, economic, and ecological considerations. For example, a hospital might have to be above the 500-year flood and a light industrial area above the 20-year level.

With these standards in mind, the next step in the merits planning process is the identification and evaluation of the options for managing the flood-prone areas as specified in a floodplain management study. From the options in the management study, a floodplain management plan is to be selected that sets out the "best combination of options available for dealing with the problem."

In reality, local governments that have completed the merits planning process have combined various elements of the process. For example, the flood study and floodplain management study are often combined. In other instances, councils have used data from flood studies conducted prior to the policy and simply updated plans to comply with the new policy. The selection of the flood standard has been more of a formality than originally anticipated in that most local councils have opted for a uniform 1-in-100 year standard, apparently on legal advice that it is a defensible state-of-the art standard. Nonetheless, there are a few examples where local councils have chosen lower standards in order to facilitate increased development.

Administrative responsibilities

The state's Water Resources Department (formerly the Water Resources Commission) and Department of Public Works have prime responsibility for state administration of the flood policy.[19] The Public Works jurisdiction consists of those areas below the riverine tidal limit while the Water Resources jurisdiction consists of those areas above this limit. In New South Wales, the tidal influence reaches many kilometers from the entrances of various rivers. For Sydney, the tidal limit is not administratively workable because it has little relationship to local government boundaries. However, the two state agencies have agreed to geographic areas of responsibility that generally follow local government boundaries, and the formal delineation of jurisdiction has not proven to be an administrative barrier for state agencies.

These two agencies are very different. Public Works is a large agency that is well funded and is politically well connected. The agency jurisdiction includes most of the highly populated areas of the state and has a strong urban focus. Public Works personnel deal with local councils that have strong engineering and planning departments. The Water Resources jurisdiction is more rural, and the department has historically been principally involved in fostering irrigation development. Many of the client councils of the Water Resources Department are small and have few professional staff members.

The regional offices of the departments of Public Works and Water Resources deal directly with local governments. The Department of Public Works oversees six regions, on which the head office in Sydney has in recent years gradually increased its grip. The head office maintains close contact on floodplain management issues within the regions as illustrated by the fact that headquarters personnel sometimes sit on local floodplain management committees. This is facilitated by the fact that there are relatively few councils within the Department of Public Works jurisdiction. In contrast, the Water Resources Department has nine regions throughout the state and numerous local councils within their purview. As a consequence, regional offices in recent years have gained increased autonomy from the department's head office.

Other state agencies also have role in flood and hazard management. By virtue of their roles in oversight of local planning and environmental management, the departments of Planning, Local Government, and Environment are potentially relevant players. For the most part, however, these agencies have taken a back seat with respect to flood hazards. Legislation that gives the State Emergency Services responsibilities for hazards management and assessment also provides a potential role for that agency. However, their focus to date has been on emergency response and hazardous materials planning.

Local government involvement

Local governments have prime responsibility for floodplain management. Occasionally, the planning and engineering personnel of a local government have opposing ideas about floodplain management. For example, the engineering department may see the solution to a flood problem as an engineering one while the planners, naturally enough, may prefer land-use or building controls. Generally within a given local government, either the planning or the engineering department takes the lead. In some cases, a professional "flood mitigation" engineer is employed by councils specifically for floodplain management planning.

Only the larger local governments have well-developed planning and engineering sections, and only some of them have dedicated flood specialists. Smaller councils, typically located in rural areas, do not have the in-house expertise necessary for developing a floodplain management plan. They will have, or share with others, an engineer whose tasks will be connected principally with roads, water supply, and sewage. For specialists, they depend on consultants or technical assistance from state government agencies.

THE CHALLENGE OF MOVING TOWARD COOPERATIVE POLICIES

From the time of Macquarie government pronouncements on floods, there has been a strong emphasis in New South Wales on the need for appropriate land-use policy for minimizing flood losses. But, floods and land-use control in Australia are intensely political. Historically, the governmental response has been one of funding structural works to protect communities from flood hazards – thereby enabling growth. Based on experience with damaging floods since the 1950s, state officials have sought different pathways to address flood hazards. One route was the highly prescriptive effort to avoid development in flood-prone areas. This engendered considerable backlash, resulting in the search for alternative regulatory regimes. The result is the cooperative, merits flood policy that provides a more flexible approach to guiding floodplain management. Although this policy has been in place for nearly a decade and it proved popular among local officials, the policy presents unresolved issues. These are briefly considered in what follows.

Sustainable policies?

Given the history of flood policy in New South Wales, a central issue is the long-term viability of a merits approach to flooding and environmental management. In particular, in an era of increased state government involvement in development decisions and increased concern about protecting environmentally sensitive areas, it is arguable as to whether the merits policy can withstand serious challenges. At issue are the political logic and sustainability of a cooperative and flexible approach.

The flexibility of the merits approach – giving it great political sustainability – may also prove to be its major weakness on two counts. One area of potential weakness is the lack of clear objectives against which to gauge success. Balancing social, economic, and ecological concerns is not a clear objective, and leaves the policy open to potential manipulation. Vigilance in applying the policy is required to stem the gradual escalation over time of flood damage potential, particularly for urban areas, while also giving weight to social and economic issues. Another area of potential weakness concerns finding the right mix of approaches to cope with flood hazards. The merits policy leaves the door open for local councils to develop comprehensive and innovative approaches, and not simply to fall back on time-tested but sometimes failed engineering approaches to flood management. The challenge is to alter thinking so that local planners and engineers are willing to innovate.

Grappling with sustainable development

In recent years, there has been widespread interest in sustainable development for which the favored Australian concept is that of "ecologically sustainable development."[20] One manifestation of this trend is the growth of the Landcare movement involving efforts of local conservation groups to manage natural resources, which has become a formal part of federal government policy and includes attention to natural hazards.

The flood merits policy, like New South Wales environmental planning policy more generally, confronts local governments with a complex task aimed at achieving vague environmental objectives. In devising these policies, state policy-makers have ventured at least implicitly into debates over what constitutes sustainability while offering mechanisms for local determination of sustainable futures as they relate to flooding or other environmental hazards. The flood merits policy clearly comes down on the side of "sustainable development" in acknowledging that flood-prone areas are valuable resources that may warrant some forms of use. However, in leaving such judgments up to local governments, the policies beg the larger task of guiding choices about sustainability.

One can envision potential for abuse of environmental decision-making under the merits policy. In local governments with strongly pro-development forces, growth within flood-prone areas would not be stopped outright and at best would be approved subject to conditions. Whether these conditions are sufficient to protect against catastrophic losses from extreme events is an open question that is of concern to New South Wales policy-makers. In principle, such factors are to be considered as part of deciding the merits of development. In practice, it is difficult to envision such considerations taking sufficient precedence to overrule development forces. At issue, then, is how local politics shapes decisions about sustainable futures.

Improved environmental outcomes are only part of the sustainable development agenda. Equity is another key element, and on this criterion the merits policy may perform well. The flexibility of the policy provides a basis for considering the diversity of needs within a community. Yet, if local decision-making is captured by a strong set of interests it is difficult to see how equitable or balanced decisions will result from the merits policy. Again, the issue is the nature of local politics and how that plays out in local decision-making.

One strength of policy-making in New South Wales for flooding is the apparent willingness of officials to learn from experience. The merits flood policy, and the extensions of the merits principle to environmental management more generally, are illustrations of such learning – albeit in the face of strong political challenges. As the flood policy matures, it is of interest to see how such political and policy learning evolves.

NOTES

1 Quoted from Thomas J. Williamson, *Housing in Flood Prone Areas* (Melbourne: Housing Research Branch, Department of Housing and Construction, 1975), p. 5.
2 Australian Water Resources Council, *Floodplain Management in Australia* (Canberra: Australian Water Resources Council, 1992), p. 6.
3 For historical discussion see Judy Fitz-Henry, "State Approaches to Flood Mitigation," in *Proceedings of the Floodplain Management Conference* (Canberra: Australian Water Resources Council, 1981), pp. 267–286.
4 For development of local government powers see Linda Pearson, *Local Government Law in New South Wales* (Sydney: Federation Press, 1994).
5 New South Wales, Department of Local Government, "Control of the Erection of Structures in the Flood Channel of Rivers Subject to Periodic Flooding," *Circular No. 1981* (Sydney: Department of Local Government, 31 January 1958).
6 The circulars are: New South Wales, Planning and Environment Commission, "Development of Flood Prone Lands," *Circular No. 15* (Sydney: Planning and Environment Commission, 16 August 1977); New South Wales, Planning and Environment Commission, "Implementation of Circular No. 15, 'Development of Flood Prone Lands'," *Circular No. 22* (Sydney: Planning and Environment Commission, 12 April 1978); and New South Wales, Department of Environment and Planning, "Zoning of Flood Prone Land," *Circular No. 31* (Sydney: Department of Environment and Planning, 15 February 1982).
7 New South Wales, Department of Environment and Planning, "Zoning of Flood Prone Land," *Circular No. 31*, op. cit.
8 New South Wales, Planning and Environment Commission, "Implementation of Circular No. 15, 'Development of Flood Prone Lands'," *Circular No. 22*, op. cit.

9 For details see John W. Handmer, "Flood Policy Reversal in New South Wales, Australia," *Disasters* 9, no. 4 (1985): 279–285.

10 Quoted from a political leaflet by the New South Wales' Liberal Party reproduced in John W. Handmer, "Flood Policy Reversal in New South Wales, Australia," op. cit. It is worth observing that household flood insurance is not generally available in New South Wales anyway, that even the strictest regulations under the policy permitted exemptions, and that there is no evidence that flood maps per se affect property values. For the latter see: Des Lambley and Ian Cordey, "The Housing Market and Urban Flood Control Policy," in *Environmental Management: Geo-Water and Engineering Aspects*, ed. Robin N. Chowdhury and M. Sivakumar (Rotterdam: Balkema Press, 1993), pp. 447–453.

11 Quoted from the New South Wales' Floodplain Advisory Committee "Policy Summary" (Sydney: Department of Public Works, memorandum, 1985), p. 3.

12 See, for example, the workshop discussion in New South Wales, Department of Public Works *Floodplain Management Issues in the Hawkesbury–Nepean Valley*, Summary Report of Workshop (Sydney: Department of Public Works, August 1994).

13 The *Environmental Planning and Assessment Act 1979* went into force on 1 September 1980. This discussion draws upon Justice P. Stein, "Environment and Planning: The Great Leap Backwards," *Law Society Journal* 24, no. 4 (1986): 62–65.

14 See Zada Lipman, *Environmental Law and Local Government in New South Wales* (Sydney: Federation Press, 1991), p. 27.

15 See commentary by Justice Stein in Zada Lipman, *Environmental Law and Local Government in New South Wales*, pp. 119–132.

16 For further elaboration see, among other sources: Linda Pearson, *Local Government Law in New South Wales*, op. cit., pp. 167–172; and Zada Lipman, *Environmental Law and Local Government in New South Wales*, op. cit.

17 These conflicts are discussed in Robert Fowler, "Environmental Law in Australia," in *Negotiating Water*, eds J. Handmer, A.H.J. Dorcey, and D.I. Smith (Canberra: Centre for Resource and Environmental Studies, Australian National University, 1991), pp. 73–91.

18 New South Wales Government, *Floodplain Development Manual* (Sydney: Department of Public Works, New South Wales Government, 1986). At the time of our study, the manual was undergoing revisions that include a name change to the *Floodplain Management Manual.*

19 After the completion of our research, the Water Resources Department and the flood, water, estuaries, and coastal management programs of the Department of Public Works were folded into a newly created Department of Land and Water Conservation. This department, which also incorporated other state resource management programs, was created to foster a more integrated approach to land and water management. See: Department of Land and Water Conservation, "The Telegraph Merger, Information Update No. 1," (Sydney: Department of Land and Water Conservation, 18 May 1995).

20 See: *National Strategy for Ecologically Sustainable Development* (Canberra: Australian Government Printing Service, December 1992).

Part II

THE POLICIES IN PRACTICE

INTRODUCTION

The rich set of policy histories discussed in Part I evidence the difficulties that policy-makers in each setting confronted in finding workable intergovernmental regimes for environmental management. The outcomes of these struggles are very different. Florida's policy-makers pursued a highly coercive and prescriptive approach. Policy-makers in New Zealand and New South Wales pursued co-operative intergovernmental approaches. As discussed in the concluding sections of the preceding three chapters, each of the policy innovations leaves a number of unresolved issues.

As they have evolved in New Zealand and in New South Wales, cooperative intergovernmental approaches to environmental management appear to have much promise. This stems from the potential for a true governmental partner-ship, local determination of appropriate forms of environmental regulation as it relates to land-use and development controls, and the political appeal of co-operative regimes to local governments. Despite this promise, basic questions remain about the ability to bring about the desired intergovernmental cooper-ation and about the efficacy of cooperative regimes.

Part II of the book examines the promise of cooperative intergovernmental regimes, based on careful analysis of the experience with cooperative regimes and how it compares with experience under the coercive regime. This assessment entails answering questions about policy implementation and outcomes: What is involved in bringing about cooperative intergovernmental partnerships? Can regional governments play a noteworthy role, given their precarious standing as governmental entities? Do local governments respond differently in preparing plans under cooperative and coercive regimes? Do cooperative policies enhance local government efforts to manage the environment? And, what are the outcomes of cooperative environmental policies? These are the central questions addressed respectively in the five chapters that comprise Part II of the book.

5

POLICY TO IMPLEMENTATION

Announcing a new policy direction is but a small step in the long chain of needed actions to bring the policy to fruition. Legislation or other authorizing actions need to be undertaken. Relevant administrative rules or guidelines need to be prepared. Governmental agencies charged with carrying out the policy need to be brought on board by securing their attention to the new policy. And, relevant government agencies need to carry out actions in support of the policy. At each step there is room for error, inevitably leading to at least some divergence between policy intent and the reality of the actions of relevant agencies. The effectiveness of environmental policies can be greatly affected by the slippage between what policies are intended to accomplish and how those intentions are translated into day-to-day actions. Of particular interest in this chapter is the experience in putting cooperative approaches to environmental management into practice.

The literature is replete with case studies of implementation gaps. One common culprit is poorly framed mandates – policies that do not adequately signal intent, guide agency actions to implement the policy, or provide sufficient incentives or other inducements for agencies to carry out necessary actions.[1] Additional problems arise when multiple agencies share responsibility for policy implementation. Conflicts over jurisdiction, concern about turf, and differences in policy interpretations can lead to variation across agencies in implementation effort and style. Variability and disjunction in policy implementation are the norm, not the exception.

Text books teach us that government agencies are the administrative arm of legislatures. However, a good deal of research on public bureaucracies shows that government agencies take additional cues from other sources.[2] Conflicting demands arise from executive and legislative principals, program beneficiaries and other interested groups that are affected by the policy, collaborating or competing organizations, and agency staff who have careers to consider and an organizational culture to maintain. As a result of these conflicting demands, agency officials must choose how much effort to put into implementing a given policy.

An understanding of these difficulties draws attention to three sets of considerations for studying the translation of policies into practice. First is policy design: does the policy design incorporate provisions that adequately signal the intent of policy-makers? Second is policy implementation: how strong are the implementation efforts of state or other agencies that have been charged with overseeing implementation? Third is the perception of policy change: to what extent do local government officials perceive a change in intergovernmental relationships?

These three considerations in assessing the translation of policy into practice are addressed in this chapter. Our primary focus is on the cooperative policies in New South Wales and New Zealand and the actions of relevant state or central government agencies in carrying out those policies. Although cooperative policies appear to have much promise in overcoming the heavy-handed approach of coercive mandates, basic questions remain about the ability to bring about the desired intergovernmental cooperation. When partners have a history of mistrust and differences over policy objectives, as was particularly true in New South Wales under the prior policy, intergovernmental cooperation cannot simply be mandated. Policies must be designed to provide incentives for cooperation. Government agencies responsible for implementing the policies must reflect that philosophy in their day-to-day dealings with lower-level governments. These governments in turn must perceive a more facilitative implementation style. Given the difficulties of bringing these conditions about, one can expect divergence between policy intent and the reality of cooperative intergovernmental policies.

An important conceptual point to keep in mind is that in this chapter we are only considering the behavior of state or central governmental agencies and their regional offices in implementing the policies under consideration. These government agency actions are important linkages in intergovernmental implementation because they establish the realities of policy as viewed from the perspective of local governments.

COMPARING POLICIES

Intergovernmental policies accomplish two tasks. Through various policy prescriptions they set forth desired behaviors of lower-level governments in developing and carrying out programs that are consistent with higher-level policy goals. Some of the prescriptions relate to the process of implementing the policy – what we label procedural prescription. Other prescriptions relate to the desired policy outcomes – what we label substantive prescription. Policies also set forth different instruments government agencies employ in helping to bring about desired governmental actions. Chief among these are the use of either coercion or incentives to build the commitment and capacity of lower-level governments, and to secure their compliance with procedural or substantive prescriptions.

The preceding chapters depicted the coercive and prescriptive policy features of growth management in Florida, the cooperative approach to resource and environmental management in New Zealand, and the cooperative overlay of the merits flood policy in New South Wales. The key distinctions in the policies are evident from a comparison of these features, as shown in Table 5.1. Of interest in this chapter are the signals that the choice of policy features establish for actions to be taken by agencies charged with implementing the policies. The discussion in this section of the chapter revisits key distinctions in policy prescription and compliance features.

Comparing policy prescription

The policies in Florida and New Zealand contain a high degree of procedural specification of planning requirements for local governments. However, the nature of the prescription is very different. Florida's growth management procedures detail requirements for policy elements, consistency of local regulations with local, regional, and state policies, and supporting documentation. The process for plan development and review is also highly prescribed. New Zealand's *Resource Management Act* also details process provisions concerning preparation of policy statements by regional councils and regulatory plans by

Table 5.1 Intergovernmental policies for environmental management

| Components | Environmental management policies | | |
	Florida's Growth Management Program	New Zealand's Resource Management Act	New South Wales' Flood "Merits" Policy
Time period	1985 to present	1991 to present	1985 to present
Intergovernmental policy approach	Prescriptive and coercive	Cooperative	Cooperative
Policy prescriptions Procedural Substantive	Restrictive and extensive	Extensive Flexible and extensive	Flexible Flexible but limited
Sanctions (*vis-à-vis* local governments)	Strong	Weak	Weak
Incentives Grants and subsidies Technical assistance	High High	Limited Moderate	High High (flood studies)
Sustainability concept	Aspects of sustainable development	Sustainable management of resources	Ecologically sustainable development

district councils. However, the specification of content is different from that prescribed in Florida. The latter details plan elements to be included while New Zealand's legislation specifies the types of effects or outcomes that may be considered (e.g., discharges of contaminants into or onto land, air or water; the avoidance or mitigation of natural hazards). Local governments in New Zealand are not required to have specific plan elements (except for coastal marine areas) or regulatory provisions. Several forms of consistency among local policies, and regional and state (national policies) are sought in both Florida and New Zealand. But, again there is a difference. New Zealand specifies that local plans must not be inconsistent with regional and national policies; providing greater latitude in local policy development than specifying detailed consistency requirements.

The flood policy of New South Wales is the least prescriptive of the three sets of policies. Local council development of floodplain management plans is not required, although they are strongly encouraged to produce such plans. There are no deadlines for local government plan preparation. Guidelines for developing floodplain management plans set forth an orderly recommended process for plan and policy development, including public participation components and mechanisms to ensure consistency with state policy. The guidelines emphasize various issues that councils should consider when developing floodplain management plans, rather than prescribing particular elements of plans or types of regulations.

Comparing compliance features

There are sharp distinctions in the compliance features of the policies. Florida's growth management legislation relies heavily on coercive features while also incorporating some of the incentives found in more cooperative policies. New Zealand's *Resource Management Act* contains review mechanisms for inducing local government procedural compliance and incentives for development of regional and local policy statements and plans. Also looming in the background in New Zealand is the threat of reorganizing local government functions; a possibility that puts regional governments in a precarious role. Procedural compliance with the flood policy in New South Wales is induced primarily through the incentives provided by conditioning eligibility for flood-control grants and the immunity of local council liability for future flood losses on following the merits-based planning process.

Florida's growth management legislation contains extensive sanctions for failure of local governments to adhere to the procedural provisions or planning deadlines. The legislation establishes severe sanctions for local governments that failed to plan, including the potential withholding of 1/365th of state revenue-sharing funds for each day a plan was late. The state also has explicit powers to take over local planning functions if necessary. New Zealand's *Resource Management Act* relies on a decentralized review process to ensure compliance with planning provisions. (Coastal policy plans are subject to direct approval by the Minister for Conservation.) The flood policy in New South Wales entails less direct state leverage to compel procedural compliance, provided under broader statutes governing state oversight of local planning functions.

All three policies entail incentives for local government development of plans or policy statements. These include grants to local or regional governments for conducting relevant studies, preparing plans, or preparing relevant policy statements. Between 1985 and 1993, the state of Florida distributed US$36 million in planning grants among the state's 457 local jurisdictions. Over the same period, Commonwealth and state funding in New South Wales for flood studies and floodplain-management planning undertaken by local governments was approximately AUD$90 million. The New Zealand Ministry for the Environment distributed some NZ$5 million per year to support studies and planning related to resource and environmental management during the first three years following enactment of the *Resource Management Act.*

One key distinction in incentives for local government participation is the way in which the different policies use assignment of liability as an inducement for policy compliance. In New South Wales, the immunity of local governments from legal liability for flood losses is used as an incentive for compliance, whereas in Florida the authorization of citizen lawsuits is used as one means of coercing compliance. State officials involved in implementing the flood policy in New South Wales see the immunity provisions as a powerful incentive for local governments to adhere to the procedural prescriptions of the merits policy. Policy-makers in New Zealand apparently considered a similar provision as an incentive for hazards planning, but concluded that immunity was unnecessary.[3] In keeping with the coercive nature of the Florida mandate, Florida takes the opposite approach to New South Wales by legitimizing citizen lawsuits against local governments that fail to plan in a manner consistent with growth management provisions.

In summary, these policies embody key features of coercive (Florida) and co-operative (New South Wales and New Zealand) intergovernmental mandates. The Florida legislation is distinguished by the detail of local plan prescription and relatively strong coercive intergovernmental components. Yet, it also contains positive inducements for local governments to undertake the required policy planning. New Zealand's *Resource Management Act* is distinguished by procedural emphasis on local government consideration of effects and a policy adoption process to help ensure that all parties have a strong voice in policy development. The flood policy of New South Wales is distinguished by strong reliance on incentives for inducing floodplain-management planning by local governments.

PUTTING POLICY INTO PRACTICE

The translation of policy into practice occurs through the actions of the government agencies charged with implementing the policy. In making this translation, government agencies are buffeted by a variety of forces from above – in the form of multiple policy mandates and oversight authorities to deal with – and from below – in the form of diverse demands from different constituency groups. As a result of conflicting demands, agency officials must choose how much effort to put into implementing a given policy or aspects of that policy. This leads to consideration of the effort that relevant agencies put into implementation.

Of particular relevance is the character of agency dealings with local governments – what we label implementation style. This can be conceptualized along a continuum ranging from a formal, legalistic style to an informal, facilitative style. Cooperative intergovernmental policies should translate into a facilitative style marked by informal dealings with local governments, flexible interpretations of policy guidance, and emphasis on reaching policy goals. Coercive intergovernmental policies should translate into a more formal, legalistic style marked by formal communications with local governments, more rigid interpretations of policy provisions, and emphasis on stricter compliance with guidelines. Failure of government agencies to translate the different policies into appropriate implementation styles would clearly evidence a substantial implementation gap.

Comparing agency style and effort across settings

We can get at these issues by comparing the agency implementation style and effort for each of the policies we consider. This summary comparison is shown in Table 5.2 for the state and central governmental agencies that are involved in policy implementation. This provides an overview of relationships among implementation style, staffing levels, and provision of assistance to local governments.

The most striking aspects of these data are the differences in agency imple-mentation style. On the scale of implementation style ranging from one (formal, legalistic) to seven (informal, facilitative), Florida's Department of Community Affairs clearly stands out as employing a more formal and legalistic approach to dealing with local governments.[4] In contrast, the relevant New South Wales and New Zealand agencies have adopted a more informal, facilitative style. These findings show that agency dealings with local governments, at least at the head-quarters level, are consistent with the relevant mandate philosophies.

Also noteworthy are the differences in implementation effort among agencies charged with policy implementation. These comparisons are complicated by incomplete data and difficulty in establishing comparable measures of effort. The most comparable measure is the ratio of the number of agency staff to the number of local governments within an agency's purview. The New South Wales' Department of Public Works has the largest staff and the least number of jurisdictions with which to work. The remaining agencies have roughly comparable ratios of staff to serviced jurisdictions. The other indicators of

Table 5.2 Agency implementation style and effort

| | Implementation style[a] | Effort Measures | | |
		Staffing ratio[b]	Technical assistance index[c]	Annual hours of assistance[d]
Florida				
Department of Community Affairs	3.3	.11	—[e]	—
New South Wales				
Department of Public Works	6.7	.54	3.0	74
Water Resources Department	5.0	.08	1.2	16
New Zealand				
Ministry for the Environment	5.5	.12	4.2	—
Department of Conservation	5.0	.14	2.3	—

Sources: Compiled by authors from surveys of relevant agencies
Notes:

[a] Rating for headquarters' staff only. Lower scores indicate a formal, legalistic style while higher scores indicate an informal, facilitative style. The index is based on the mean rating of three items (1 to 7 scale for each item): agency communications with local governments (written to less formal), interpretation of policy guidance (as strict rules to flexible interpretation), and substantive emphasis (strict adherence to policy goals).

[b] Ratio of headquarters staffing to number of local governments served by the agency. Staffing data are for relevant divisions of the agencies. Data for New South Wales are for divisions only dealing with floodplain management. Data for New Zealand are for relevant agency divisions charged with implementing the *Resource Management Act*.

[c] Index computed for each relevant jurisdiction as a sum of a set of agency actions (each 1 = yes, 0 = no; maximum score of six): responded to questions, reviewed plan or policy, distributed guidance materials, distributed example plan or policy, telephone consultation, on-site technical assistance. Cell entries are mean values among relevant local jurisdictions for which regional office data are available.

[d] Mean number of hours of technical assistance provided for the most recent year per local government within agency purview.

[e] Data are not available.

implementation effort are the extent of technical assistance and the hours of assistance provided to local jurisdictions.[5]

Not surprisingly, more staff make possible higher levels of technical assistance and greater hours of assistance. Yet, there does not appear to be a simple relationship between staffing levels and agency implementation style. The ratio for Florida's Department of Community Affairs is comparable to those of the New Zealand agencies, but the Florida agency has a much more legalistic implementation style. Although a minimum threshold of staffing levels seems critical for fostering a facilitative implementation style, these data suggest more is involved in shaping implementation style than staffing ratios.

Considering variation within settings

Differences in degree of agency effort and style within New South Wales and within New Zealand are also evident from Table 5.2. Although the differences are not as striking as those across settings, the differences suggest that the co-operative mandates are not being embraced with equal fervor by agencies. This variation can be explained with respect to different levels of agency involvement in development of the policies, different agency philosophies about their roles, and different agency cultures.

The largest differences are within New South Wales where two agencies have primary responsibilities for implementing the merits floodplain management policy. (As noted in Chapter 4, the Department of Planning also has a potentially important role, but it has chosen to play a secondary role with respect to flood-plain management.) These agencies divide their flood-related responsibilities so that the Department of Public Works deals with governments in tidal areas (coastal and estuaries) and the Water Resources Department deals with inland councils. The data in Table 5.2 show that Public Works has devoted nearly three times as many staff to the floodplain management program, provides nearly three times the level of assistance to the average jurisdiction within its purview, and provides on average more than four times as many hours of assistance as does the Water Resources Department. Both agencies report having adopted facilitative implementation styles, with Public Works reporting a somewhat more facilitative approach.

The differences in effort in New South Wales can be explained by several factors. Senior personnel from Public Works were actively involved in developing the merits policy, and they took the lead in developing the guidance materials for the policy. The policy emphasis on urban flooding, other than rural flood losses, increases the relevance of Public Works.[6] In addition, Public Works had a long history of working with local governments in planning flood control and other public works. As part of that involvement, agency officials recognized early on (based in part on observing floodplain management in the United States) that new approaches had to be developed to manage flood risks. This led to an agency culture supportive of floodplain management and so-called non-structural solutions to flood hazards.

Officials from the Water Resources Department, however, had less involve-ment in developing the merits flood policy. Some departmental officials appear to read the urban emphasis of the policy as signaling a less important role for their agency. The more restrained approach is also reflective of a different agency history. Water Resources has a long history of working with farmers in providing advice about water resources and related issues, rather than working with local governments. Although floods were of concern to farmers and the agency, the merits policy added a very different set of demands for the agency.

Differences in agency effort are also evident for the relevant New Zealand agencies. These differences are explained by a set of historical factors and by differences in agency culture. Ministry for the Environment personnel were central players, along with Treasury officials, in developing the *Resource Management Act*. Building on the strong roles specified in prior environmental statutes, the resource management legislation designated the Ministry for the Environment as the lead agency for implementing the policy, except for coastal provisions designated to the Minister for Conservation. The more focused role for the Department of Conservation resulted in that agency's attention being drawn to development of a national coastal policy, which took priority over other activities. As a consequence, it is not surprising to see the Department of Conservation devoting fewer resources to technical assistance.

The differences in agency implementation style between the New Zealand agencies are in the direction that we expect, but are not as great as anticipated. Because the role of the Ministry for the Environment evolved during policy enactment into that of a broker between Treasury and environmental forces, a relatively facilitative agency implementation style could be expected. However, it appears that the limited number of staff is a constraint. The somewhat less flexible approach of the Department of Conservation is consistent with the agency's conservation-oriented culture, and responsibilities as an advocate for conservation.

EXAMINING REGIONAL VARIATION IN IMPLEMENTATION STYLES

The preceding comparison of agency styles across settings shows that agencies are adopting implementation styles that are consistent with the intent of relevant policy mandates. Yet, the differences in implementation styles among them suggest that the aggregate comparisons mask underlying variation. Indeed, it would be surprising if different agencies or regional offices of the same agency responded with equal fervor or adopted the same implementation style. As is true of their parent agencies, regional offices must deal with conflicting demands that cannot always be addressed in the same fashion.

Analyzing regional variation has several advantages. It expands the number of cases sufficiently to permit statistical analyses. These analyses in turn provide a basis for commenting about the agency and situational factors that affect implementation styles. Because the regional offices are active in day-to-day dealings with local governments, this level of analysis also provides a more realistic portrayal of policy in action. The main disadvantage of considering variation among regional offices is that we are forced to restrict attention to New South Wales. The data for the other settings are too incomplete to provide a sound basis for similar analyses.

The regional office data for New South Wales consist of information collected from five of the six regional offices of the Department of Public Works and eight of the nine regional offices of the Water Resources Department. (Both of the non-responding offices are in areas with relatively low flood risks.) Lead personnel in each office filled out questionnaires providing factual information about staffing and the extent of contact with local governments within their agency purview. The questionnaires also solicited responses to a series of questions asking about perceptions of the regional office's capabilities, commitment, and approach to dealing with local governments. In order to obtain measures of the seriousness of the hazards problem and demands for its solution within the region, we computed median scores for relevant variables for the local governments within each region. These data come from our survey of local governments in New South Wales.

Variation in style among regional offices

Several key differences emerge in looking at the implementation styles of regional offices of state agencies in New South Wales.[7] First, there is considerable variation in regional office implementation styles. On the scale of one (legal, formalistic) to seven (informal, facilitative), the implementation scores for regional offices ranged from 2.3 to 5.7, with a mean score of 3.7. Second, there are noteworthy differences in mean scores between agencies. The mean implementation style score for regional offices of the Department of Public Works is 4.3 while the corresponding score for regional offices of the Water Resources Department is 3.4.[8] Third, there is a disjunction between the more facilitative headquarters scores for both agencies and the mean regional office implementation style scores.

The variation among regional offices in implementation styles and disjunction from headquarters evidence common difficulties in fully translating policy from headquarters to those on the front lines of policy delivery.[9] It is not surprising that the translation of policy to practice has been more effective for the Public Works than for the Water Resources. As noted in the discussion of implementation effort, the leadership of the Department of Public Works has strongly embraced the policy, and the agency has a stronger historical record in working with local governments. The fewer number of regional offices for Public Works permits more frequent interactions with headquarters personnel and better communication. Public Works also devotes more staff in regional offices to work on floodplain management relative to the number of jurisdictions than does Water Resources. A common lament of regional office personnel of the Water Resources Department was the lack of adequate staffing for carrying out flood-related assistance.

As we would expect, the level of technical assistance is higher on average in those offices reporting a facilitative approach than those reporting a more formal implementation style. The availability of such assistance is facilitated by greater levels of staffing relative to the number of jurisdictions in the region.[10] There are also noteworthy differences between the two groups of offices in perceptions of regional office capacity and commitment.

Explaining variation in implementation style

We developed a statistical model to identify factors that account for the variation in implementation styles among regional offices of state agencies in New South Wales. The potentially relevant factors are those that are internal to the agency (or at least reflect decisions made by agency officials) and those situational factors that are beyond the immediate control of agency personnel. The internal factors include the commitment and capacity of regional offices, whether the office is part of Public Works or Water Resources, and the number of local governments within the purview of the regional office. The situational factors reflect a mix of indicators of willingness of local personnel to engage in hazard management, demands from neighborhood groups to address flood hazards, and severity of risk.

The statistical evidence shows that more formal implementation styles are fostered by having to deal with larger numbers of jurisdictions or by having greater proportions of jurisdictions with prior losses from disasters.[11] More facilitative implementation styles are fostered by increased agency commitment and capacity, greater commitment of local governments to floodplain management, and increased demands by neighborhood groups to address flood hazards. Controlling for other factors, regional offices of the Department of Public Works have on average more facilitative styles than the regional offices of the Water Resources Department.

The most noteworthy finding is the influence of agency commitment and capacity, which accounts for one-third of the variation in implementation style. This finding is consistent with findings from prior research that also showed a conditional effect of agency commitment and capacity on agency implementation style.[12] The small number of cases under study prohibits use of regression models for sorting out the details of this interaction.

One way of sorting out the effect on implementation style of differing levels of agency commitment and capacity is to compare the median scores for implementation styles for regional offices reporting different levels of these factors. Table 5.3 presents this comparison, which involves no statistical controls. Cell entries are implementation style scores for which our interest is in the degree to which regional offices pursue a facilitative style in dealing with local governments within their purview. Comparisons of commitment and capacity are based on splitting the cases at the median values of all observations.

These results suggest that when agency commitment and capacity are both high, the combined effect enables an informal, facilitative implementation style. But when commitment is high and capacity is low, the combined effect stimulates a more formal, legalistic style. Indeed, the median implementation style for this combination of factors is even more formal (i.e., has a lower score) than when both commitment and capacity are low. High commitment presumably leads to pressures on agency personnel to pursue the cooperative policy, but the lack of corresponding capacity limits ability to carry through with

Table 5.3 Effect of agency commitment and capacity on implementation style
(New South Wales)

Levels of Regional Office[a]		Regional Office Statistics	
Commitment	*Capacity*	*Implementation style*[b]	*Number of cases*
Low	Low	3.33	7
High	Low	2.67	2
Low	High	4.33	1
High	High	5.67	3

Sources: Compiled by authors from surveys of relevant agencies

Notes:

[a] Data are split so that regional office values at the median or below are considered low and those above the median are considered high. Commitment and capacity measures based on respondent ratings from agency surveys.

[b] Median of the implementation style scores for those offices with the specified combination of commitment and capacity. Higher scores indicate a more informal, facilitative implementation style.

facilitative actions. The relatively high score for the combination of high agency capacity and weak commitment suggests that capacity is more important than commitment in bringing about a facilitative implementation style. However, this finding is tempered by the fact that only one observation fits that profile.

In short, these findings suggest that both commitment and capacity need to be strong for agencies to adopt a facilitative implementation style. This is because higher agency commitment and capacity provide greater agency freedom and confidence in pursuing more cooperative approaches. To the extent that facilitative approaches are desired, increasing agency commitment alone may not be sufficient. Indeed, these findings suggest that a high level of commitment, without a commensurate level of capacity, leads to adoption of a more formal, legalistic implementation style.

PERCEPTIONS OF LOCAL GOVERNMENT OFFICIALS

The adoption of facilitative implementation styles by state (or national) government agencies means little unless local government officials sense a co-operative approach. If they perceive state or central government officials as being hard-nosed, local officials may respond in kind or possibly resist by not complying with the policy.[13] If they perceive a more facilitative approach, they may be more likely to embrace the cooperative policy and collaborate with higher-level officials. We turn in this section to the perceptions that local government officials have about their dealings with state agencies in New South Wales and central government agencies in New Zealand.

We gauge these perceptions from the surveys of local government that we conducted in each setting. The New Zealand data consist of a separate set of responses for district councils and for regional councils – the two units of local government that are involved in complying with the planning provisions of the *Resource Management Act*. The data for New South Wales consist of responses for local councils. Respondents were asked to rate various aspects of state (or in New Zealand, central government) approaches to dealing with local governments before and after cooperative policies came into effect. Changes in scores for corresponding items provide a sense of differing perceptions over time. The obvious caveat is that we are working with retrospective data for the assessment of relationships prior to the policy changes (some eight years in New South Wales and two years in New Zealand). Although memories may be distorted, we are reassured that many of the respondents held similar career positions prior to the introduction of the newer policies.[14]

Given the shift toward cooperative intergovernmental policies in New South Wales and in New Zealand, change toward more facilitative approaches is expected in each setting. The changes should be strongest in New South Wales because the preceding government policy was a highly coercive policy involving strong state government regulation, and because local governments, at the time of our survey, had nearly eight years' experience with the merits cooperative policy. Nonetheless, we expect variation in the perceptions of local officials in New South Wales. Not all localities have participated in the planning process set forth under the merits policy, and, as shown earlier, there is variation in the way state agencies have translated the policy into day-to-day dealings with local governments. The changes in New Zealand are expected to be more muted because the cooperative policy is more recent, and because the degree of change in intergovernmental relationships under the act is not as sharp.

Table 5.4 shows the perceptions of local government officials of change for various aspects of higher-level agency dealings before and after cooperative policies went into effect. Each item is rated on a scale of one to seven with the end-points of each scale noted in parentheses after the item. Higher scores indicate perceptions of more facilitative agency dealings with local governments.

For all measures in New South Wales and key measures in New Zealand, there is a noteworthy change in perceptions over time in the direction of more cooperation. The greatest change in both settings is for the item measuring substantive policy emphasis – whether the government's emphasis is on planning standards and rules, or on planning goals and outcomes. The strong perception of an emphasis on goals and outcomes clearly indicates that the central thrust of cooperative policies is being understood by officials in local governments.

The New Zealand data concerning government cooperation and persuasion show the least change. Because there was an inconsistent history of the use of sanctions and incentives for environmental planning in New Zealand, it is not surprising to see little change in the level of intergovernmental persuasion. Indeed, a number of respondents saw this as an irrelevant item. The perception of less extensive government cooperation by regional officials in New Zealand

Table 5.4 Perceptions of changing implementation styles

Local governments/Scale item[b]	Perception of agency style[a]		
	Before policy	After policy	Difference[c]
New South Wales local councils			
Substantive emphasis (rules vs. goals)	2.67	5.24	2.57 ***
Government cooperation (limited vs. extensive)	3.47	4.78	1.31 ***
Persuasion (sanctions vs. incentives)	3.95	4.94	.99 ***
New Zealand district councils			
Substantive emphasis (rules vs. goals)	2.61	5.43	2.82 ***
Government cooperation (limited vs. extensive)	3.98	4.37	.39 **
Persuasion (sanctions vs. incentives)	3.85	3.67	−.18
New Zealand regional councils[d]			
Substantive emphasis (rules vs. goals)	3.20	5.33	2.13 ***
Persuasion (sanctions vs. incentives)	4.92	4.46	−.46
Government cooperation (limited vs. extensive)	4.47	3.60	−.87 *

Sources: Compiled by authors from surveys of local governments
Notes:
  * p <.1 ** p <.05 *** p <.01

a Cell entries are mean values for perceptions that local government officials have of different aspects of the implementation styles of relevant state (or national) agencies. Before and after scores consist of separate ratings by respondents of each item as part of local government questionnaires. Higher scores indicate perceptions of an implementation style that is informal and facilitative.

b Each item is based on a 1 to 7 scale with endpoints anchored by the labels in parentheses. Responses were obtained from 127 local councils in New South Wales. For New Zealand responses were obtained from fifty-nine district councils, sixteen regional councils, and four unitary authorities.

c Difference in means before and after policy for which positive differences indicated changes in the direction of a perception of more informal, facilitative agency styles. Statistical tests are t-tests of paired differences for which the pairs consist of corresponding before and after ratings. One-tailed p-values are reported.

d New Zealand regional councils and unitary authorities.

might be explained by the fact that at the time of our data collection, the regional districts were grappling with lead roles in implementing the *Resource Management Act*. Some officials felt there was insufficient guidance from central government agencies about the act's provisions, and that they were being asked to undertake extensive policy development without adequate staff or assistance. This view is reinforced in a report prepared for the New Zealand Business

Roundtable about policy implementation: "[T]here is some justification in an oft-repeated criticism, namely that having passed a novel and highly sophisticated piece of legislation, central government has walked away and left much of the hard work to be done by local government."[15]

As expected, respondents from New South Wales report noteworthy changes in perceptions of policy approaches over time for a range of indicators. The changes in perception among those local councils who actively participated in the policy were greater than those who chose not to undertake the suggested planning process.[16] This provides further evidence that the cooperative policy contributed to changed perceptions of intergovernmental relationships. The relationship between the implementation style adopted by regional offices of state agencies in New South Wales and changes in local government perceptions of state agency relationships is more muted. The strength of the relationship is weak for each of the indicators, and only slightly higher, when also taking into account the effort expended by regional offices.[17] It seems that councils become more aware of agency styles as the agencies put more effort into working with local governments.

One explanation for the modest relationships between regional office implementation style and changes in local perceptions of these styles is that local governments also deal directly with headquarters offices. Personnel from the Department of Public Works estimate that some 50 percent of staff time by headquarters personnel is spent working directly with local governments. Personnel in the regional offices of the Water Resources Department also report direct headquarters to local government cooperation. As a consequence of involvement by headquarters personnel, the local government respondents may be basing their perceptions of agency relationships largely on their relationships at that level. We noted earlier that headquarters personnel report more facilitative styles than the mean implementation style of corresponding regional offices of state agencies. This also helps explain why there is a strong policy effect across a range of local governments who participated in the merits policy.

CONCLUSIONS

This chapter has addressed the translation of policy into day-to-day actions of relevant government agencies and resultant local government perceptions. The coercive growth management policy in Florida is reflected by a high degree of policy prescription and coercive features for planning by local governments. The cooperative intergovernmental approach of New Zealand's *Resource Management Act* incorporates prescription about policy process, but does not use coercion to gain compliance by local governments with that process. The cooperative flood policy in New South Wales is neither highly prescriptive nor highly coercive with respect to floodplain-management planning by local governments.

In examining the translation of policies into practice, differences in implementation efforts are notable among the policies as well as among different government agencies charged with implementation. The character of agency dealings with local governments is particularly important. When viewed at the headquarters level, the implementation styles of relevant government agencies are consistent with corresponding coercive and cooperative intergovernmental policies. However, gaps in translating policy into practice are especially evident among regional offices of government agencies.

We attribute these gaps to a mix of internal and external forces. The commitment and capacity of state (or national) agencies are especially important factors, for which both need to be strong in order to foster implementation styles that are informal and facilitative. High levels of these factors provide greater agency freedom and confidence in pursuing facilitative approaches. These findings suggest that a high level of agency commitment without a commensurate level of capacity leads to adoption of more formal, legalistic styles.

Looking at agency actions reveals only one side of the translation of policy into practice. Also relevant are the perceptions that local government officials have of their day-to-day dealings with government agencies, and how those relationships have changed under cooperative policies. Local government officials in New Zealand and in New South Wales perceive more collaborative relationships under the cooperative policies than under prior coercive or prescriptive policies. However, the perceptions are not uniform. The positive perceptions are greatest among officials in local governments that have made more extensive efforts to carry out the cooperative policies.

The findings of this chapter suggest that cooperative and coercive intergovernmental policies engender similar implementation problems, but cooperative policies raise additional difficulties. One common problem is gaining commitment of the state or national agencies charged with implementation to carry out policies. This can be particularly difficult for agencies that were not active in the policy development. A second, common problem is gaining uniformity in the translation of policy at the front lines of policy delivery among regional field offices. These findings also suggest that cooperative policies are more difficult to translate into practice than are coercive policies. Cooperative policies require high levels of commitment and capacity of state (or national) agencies in order to foster the facilitative styles of implementation. The design of policy mandates is important in signaling cooperative desires and structuring intergovernmental relationships.

NOTES

1 See Helen Ingram and Anne Schneider, "Improving Implementation Through Framing Smarter Statutes," *Journal of Public Policy* 10, part 1 (January–March, 1990): 67–88; Daniel A. Mazmanian and Paul Sabatier, *Implementation and Public Policy* (Glenview, IL: Scott Foresman, 1983).

2 For an overview of this research tradition see James Q. Wilson, *Bureaucracy, What Government Agencies Do and Why They Do It* (New York: Basic Books, 1989).

3 David Bewick, "Natural Hazard Mitigation Aspects of the Resource Management

Law Reform and the Reform of Building Controls," paper presented at the Third Australian/New Zealand Floodplain Management Forum, New South Wales (Wellington: Ministry for the Environment, April 1991).

4 Implementation style is an index comprised of the mean rating on a one to seven scale of each of three items: agency communications with local governments (written to less formal), interpretation of policy guidance (as strict rules to flexible interpretation), and substantive emphasis for compliance (strict adherence to policy goals). Ratings for each item were obtained from questionnaires filled out by key personnel within relevant agencies charged with implementing the policies under study. Similar data were also obtained from key personnel in regional offices of New South Wales and New Zealand agencies. The Cronbach alpha of the index of .82, computed using ratings by personnel in New South Wales regional offices, suggests a reasonable level of reliability for the index. Higher scores indicate a more informal, facilitative implementation style, while lower scores indicate a more formal, legalistic style.

5 The technical assistance index is the total number of actions taken in dealing with a given local government over the past twelve months, out of six possible activities: responding to questions; reviewing plans or policies; distributing guidance materials; distributing sample plans or policies; providing telephone consultations; or providing on site technical assistance. These were compiled from reports of regional office personnel of the extent of their contacts with individual local governments within their purview.

6 The mean risk of flooding, as reported by local respondents as part of our survey of local governments, is not statistically different between local governments within the different jurisdictions of the Department of Public Works and the Water Resources Department (difference of means for the 100-year flood threat, t-value of .55, p = .58). However, the mean 1991 population is nearly three times as large for jurisdictions of the Department of Public Works than it is for jurisdictions of the Water Resources Department (difference of means t-value of 5.28, p <.01).

7 Similar patterns in implementation styles of regional offices are also evident for central government agencies in New Zealand. The mean implementation style score for three of the four regional offices of the Ministry for the Environment is 4.67, while the corresponding score for twelve of the fourteen regional offices of the Department of Conservation is 4.04 (p = .08 for a difference of means test). Mean scores for regional offices in New Zealand are lower than their corresponding scores for headquarters offices, indicating more formal implementation styles by the field than by headquarters.

8 The p-value for a difference of means t-test for the two sets of offices is .10.

9 Research that has addressed regional variation in environmental policy implementation includes David M. Hedge, Donald C. Menzel, and George H. Williams, "Regulatory Attitudes and Behavior: The Case of Surface Mining Regulation," *Western Political Quarterly* 41, no. 2 (June 1988): 323–340; and Neal Shover, John Lynxwiler, Stephen Groce, and Donald Clelland, "Regional Variation in Regulatory Law Enforcement, The Surface Mining Control and Reclamation Act of 1977," in *Enforcing Regulation*, ed. K. Hawkins and J. Thomas (Boston: Kluwer-Nijhoff Publishing, 1984), pp. 121–145.

10 The Pearson correlation between staffing per jurisdiction and implementation style is .49 (p <.05). The corresponding correlation between staffing per jurisdiction and the mean level of assistance in the region is .69 (p <.001).

11 This entailed explaining variation in implementation style among the thirteen regional offices for New South Wales (adjusted R2 = .87, F-value = 12.18, p <.01). Positive coefficients indicate contributions to a more informal, flexible implementation style. The explanatory measures included three measures of agency characteristics for which

the standardized coefficients are as follows: interaction of agency commitment and capacity ($B = .61$, p <.01), square root of the number of jurisdictions in the region ($B = -.53$, p <.01), and lead agency (1= Public Works, 0 = Water Resources; $B = .37$, p <.01). Also included were four situational factors for which the standardized coefficients are as follows: commitment of local government agencies ($B = .36$, p <.05), constituency demands by neighborhood groups ($B = .33$, p <.1), prior losses from disasters ($B = -.53$, p <.01), and estimated current flood risk ($B = -.02$, p = .45). Details about this statistical model are reported in Peter J. May, "Can Cooperation Be Mandated? Implementing Intergovernmental Environmental Management in New South Wales and New Zealand," *Publius, The Journal of Federalism* 25, no. 1 (Winter 1995): 89–113.

12 Peter J. May, "Mandate Design and Implementation: Enhancing Implementation Efforts and Shaping Regulatory Styles," *Journal of Policy Analysis and Management* 12, no. 4 (Fall 1993): 634–663.

13 A study by Stephen Jenks suggests that such responses do not always follow since there may be mutual gains in adopting a problem-solving approach in attempting to work out differences, rather than simply let the conflict escalate. See Stephen Jenks, "County Compliance with North Carolina's Solid-Waste Mandate: A Conflict-Based Model," *Publius: The Journal of Federalism* 24, no. 2 (Spring 1994): 17–36.

14 Also complicating retrospective comparisons for New Zealand are substantial local government changes in the late 1980s, which involved local government boundary changes and the creation of newly constituted regional governments. However, many of the personnel that were staff to the prior catchment boards became staff of regional councils and many of the district council staff continued during local government changes.

15 Alan Dormer, *The Resource Management Act 1991, The Transition and Business*, prepared for the New Zealand Business Roundtable (Wellington, NZ: Business Roundtable, August 1994), p. 62.

16 The comparison is between those governments which chose not to participate or only took minimal steps (n = 48), and those governments who were currently undertaking or had already completed the planning process (n = 78). Two of the t-tests of differences of means are significant at the .05 level or less (substantive emphasis, persuasion) while the remaining item is significant at the .10 level (government cooperation).

17 The Pearson correlation between regional office implementation scores and the mean change in scores for the various local perceptions within each region ranges from .04 to .30. The correlation for the interaction of regional office effort (staff per jurisdiction) and implementation style when correlated with the mean changes in local perceptions for each region is .05 for substantive emphasis, .18 for government cooperation, and .58 for persuasion.

6

A REGIONAL
GOVERNMENT ROLE

One of the central questions in the design of environmental management regimes is the role that is granted to regional entities. Like the state agencies discussed in the previous chapter, regional entities are sometimes cast as important players in environmental management. This is the case for New Zealand's regional tier of government and in different ways for Florida's regional planning councils. Unlike state agencies, regional entities are often placed in a precarious role because they do not have the degree of authority that is vested in state agencies. Nor do they have the political base with the citizenry that local governments derive from their direct relationship in service delivery. As a consequence, the standing of regional entities is often precarious.

Despite this potential limitation, a strong case can be made for a role for regional entities in addressing natural resource and environmental issues. Properly designed regions with jurisdictional boundaries that are congruent to natural boundaries provide a basis for managing resources and environmental conditions at a meaningful scale.[1] This is the logic of catchment boards and river basin authorities that have been created for managing watersheds. Regional approaches are not limited to resource management, as they have also been employed in addressing environmental problems. This is illustrated by the creation of air pollution districts or flood control districts.

Regional approaches to dealing with resource management and environmental problems have been piecemeal, if they exist at all, in most countries. The United States is perhaps exceptional in the variety of forms of regional entities and special districts that have been created for dealing with such issues. While there have been some successes, including some of the regional authorities noted above, the legacy of regional governments for managing resources and addressing problems in the United States is largely a disappointing one.[2] The experience in Australia is more limited in attempting substate regional solutions to environmental problems, although there have been some successes with local governments forming regional authorities for addressing flood hazards. Neither of these experiences constitute a well-planned, systematic approach to a regional role in resource or environmental management.

For several reasons, New Zealand's approach in establishing a regional role is exceptional in comparison to the approaches of other countries. Unlike the creation of a plethora of *ad hoc* regions or special districts in other settings, New Zealand's policy-makers consolidated a variety of environmental and resource management functions within a country-wide set of regional governments. As such, the approach is far more comprehensive than found elsewhere with respect to both the geographic area covered and the range of environmental management responsibilities. Unlike the limited powers granted to regional authorities in other settings, the regional councils in New Zealand were charged with note-worthy planning and policy-setting responsibilities. As discussed in Chapter 3, this reflects the conscious decision by New Zealand's policy-makers to devolve decision-making to regional and local levels of government. In addition, regional councils in New Zealand have political and financial bases by virtue of having elected representation and having their own revenue sources. In short, the regional approach undertaken in New Zealand stands out as a serious effort to instill a comprehensive, rational approach to resource and environmental management.

Despite the intent, there are clear challenges to establishing an effective role for regional governments in New Zealand. One challenge is accomplishing the planning and policy-setting tasks that have been assigned to regional councils. These are new undertakings that require different ways of doing business and thinking than was the case for predecessor regional entities. A second challenge is fostering the regional and local government partnership that is sought under the *Resource Management Act*. Given a history of mistrust and conflict among these layers of government in some settings, positive intergovernmental relationships cannot be assumed from the outset. A third challenge is the viability of a regional role for environmental and resource management. A basic issue is whether or not the newly re-constituted tier of regional government can withstand criticism as problems are encountered.

Given the appeal of a regional government role on the one hand but the note-worthy challenges that face regional governments in fulfilling that role on the other hand, the regional dimension constitutes one of the less certain aspects of New Zealand's policy innovation in resource and environmental management. This chapter addresses the role of regional governments and how it seems to be evolving. At issue is the extent to which regional governments are fulfilling their promise in New Zealand's devolved, rational system of environmental management. Key elements of that promise are a sounder basis for guiding environmental planning and the prospects for regional governments in assisting local governments in their efforts to address environmental problems. This draws attention to the performance of regional governments within the context of a cooperative intergovernmental policy.

REVISITING REGIONAL GOVERNANCE

Establishing a role for regional governments in environmental management presents opportunities and challenges that differ from those presented by roles for national, state, or local governments. The desire for strong roles for national or state governments stems from both a need for uniformity in overall purpose and a political need to back government intervention with the power of the state. The logic for local government responsibility stems from basic needs for a manageable scale of government action and the desire for local responsiveness. State or national governments are not good at fixing storm drains. Local governments are not good at designing national environmental programs. The opportunities and challenges posed by establishing a role for regional governments are discussed in this section.

Logic of a regional government role

Perhaps the strongest argument for creating regional governments and vesting them with environmental responsibilities is that they can address extra-local problems. As noted earlier, this makes most sense for environmental management when the regional boundaries have a natural basis. The establishment of regional policies for such areas, which serve to guide local planning, has the potential for overcoming problems that arise when undertaking piecemeal local planning for only parts of a given ecosystem or region. As such, regional governments have the potential for providing a strong planning and coordinating mechanism for guiding local decisions.

This coordinating and facilitating planning function is one logic used in creating a regional planning role as part of Florida's growth management program.[3] The state comprehensive plan, for example, recognizes the need for intergovernmental coordination and the use of regional approaches in addressing over a dozen issue areas such as public safety, water resources, coastal marine resources, and land use. As discussed in Chapter 3 and elaborated upon later in this chapter, New Zealand's regional councils also have important regional policy preparation and coordination functions.

One aspect of regional coordination is facilitating communication both among local governments and between local governments and higher levels. Closely related to this is the provision of technical advice and other forms of assistance to local governments. It is not clear that regions necessarily have greater expertise than local governments. But under a decentralized system of government like New Zealand's, the logic is that regions will develop requisite expertise and share lessons with less capable local governments.

Regional councils can play important roles in developing regional environmental data bases and providing technical assistance to local governments, but simply thinking about regional roles in service delivery terms misses a larger debate about the distribution of governing authority. This, of course, is central to the New Zealand experience since a key element of governmental reform was to shift responsibilities to newly constituted regional (and local) councils, while reducing central government involvement.

Instability of regional governments

A weakness of regional tiers of government is that they pose potential threats to other levels of government. No matter how desirable it is to have regional involvement, there is no escaping the fact that they constitute another layer of governmental intervention. Local officials see funds going to regional organizations that could be coming to them, and they concerned with potential encroachment on local responsibilities. Rather than easing intergovernmental relations, these turf considerations have the potential for increasing conflict and fragmenting implementation.

Florida's experience in carving out a regional role illustrates these tensions. There was long-standing tension between local governments and the independently authorized and funded regional water-management districts. Partly to avoid exacerbating these conflicts, Florida's policy-makers assigned regional planning functions under the 1985 growth management legislation to a different set of regional authorities, involving regional planning councils that had weaker standing. As noted in Chapter 2, from their inception in 1972 the regional planning councils in Florida have been controversial. Even though local officials hold two-thirds of the seats on each regional council (the other third are appointed by the governor), local governments resent regional interference in local affairs, just as they resent state interference. The precariousness of the regional planning function was underscored by efforts in the 1992 legislative session to abolish regional planning councils. This eventually resulted in legislation redefining their roles in order to minimize potential interference with local planning.

The history in New Zealand with regional entities also displays instabilities in regional government roles. Through a series of acts dating from the abolishment of regional provinces in 1876, regional entities have been created, abolished, or reassigned roles many times over. By the time of local government reform in the late 1980s, a variety of regional organizations had emerged including United Councils – assigned regional planning functions, Catchment Boards – assigned water resource and soil management functions; and Metropolitan Regional Councils – assigned metropolitan planning functions.[4] When consolidating these regional functions in 1989 into the newly re-constituted regional councils, central government policy-makers reassured local officials that the new regional councils would not constitute a superior form of government to them, nor would the regional councils undertake local government functions. The current construction of the responsibilities of regional governments attempts to minimize conflict between regional and local governments.

The need for the newly created regional councils was not readily accepted, and not until the *Resource Management Act* was adopted did a strong case exist for them. Indeed, a review was undertaken a few years after the establishment of regional councils in which the need for regional councils was re-assessed.[5] The conclusion was that dismantling regional councils would undermine the workings of the resource management legislation, and would also be a politically unacceptable outcome. As a further reassurance to local governments, the functions of regional governments were restated as part of amendments to local government legislation in 1992, removing the last vestiges of non-environmental responsibilities. These steps can be viewed as a renewed commitment to a strong role for regional governments in resource and environmental management.

Although central government policy-makers were reluctant to disassemble the intergovernmental framework established under local government and resource management reforms, there are options for other governmental structures. As noted in Chapter 3, local government legislation in New Zealand gives local councils the authority to take on the dual functions of regional and local councils. This is not a simple undertaking, since it requires that they meet certain conditions and demonstrate to the Local Government Commission that they can perform regional functions as efficiently or effectively as the regional councils. As of 1995, four such unitary authorities had been created.

EXPECTATIONS AND CAPACITY OF REGIONAL GOVERNMENTS

As noted in the introduction to this chapter, one of the unique aspects of New Zealand's environmental management regime is the strong role that has been carved out for regional governments. An obvious issue is their capacity to carry out these functions. The expectations and capabilities of regional councils are addressed in this section.

Expectations of regional governments

As discussed in Chapter 3, regional councils are required by the *Resource Management Act* to prepare regional policy statements that provide an overview of resource management issues and propose policies to achieve integrated management of natural and physical resources for the relevant region. Following from the regional policy statements, regional councils can prepare regulatory regional plans, but these plans are not compulsory except for the coastal marine area. The policy planning requirements are formal functions for regional councils, in addition to others detailed in Chapter 3, that are clearly stated in the legislation. One benchmark for assessing regional government performance consists of evaluating their performance in preparing the required policy statements, recognizing that many of these were still being finalized at the time of our study.

The less explicit expectations of regional councils relate to their roles in a devolved system of environmental management. Given the restricted roles envisioned for central government agencies in addressing local problems, it is clear that regional governments were anticipated as picking up at least part of the void that remained. This includes helping to coordinate local planning,

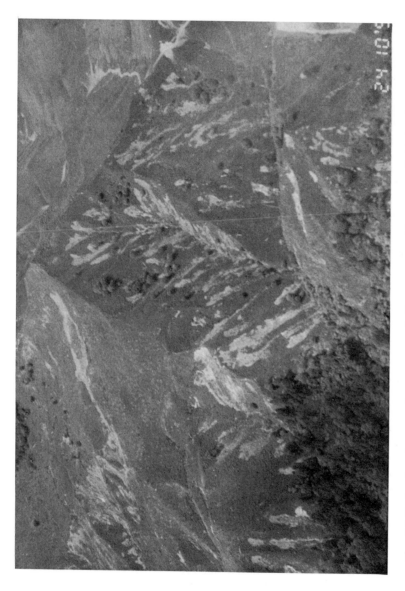

Plate 6.1 Regional environmental management challenges presented by erosion of hillsides cleared for pastoral farming in the central North Island, New Zealand (courtesy of Environment Waikato, New Zealand)

which is an explicit objective of the resource management legislation. It also includes provision of technical assistance to local governments. In these roles, regional councils are expected to be facilitators for local government planning. Thus, a second benchmark for assessing the roles of regional governments is the extent to which they are fulfilling these facilitative roles.

The extent to which regional governments were expected to fulfill roles in environmental monitoring and assessment that had previously been undertaken by central government agencies are less clear. Such information is clearly important for undertaking "effects-based" environmental management. The appropriate level of government for undertaking the necessary database development appears to be an open issue. In discussing these needs, the Parliamentary Commissioner for the Environment, Dr. Helen Hughes evaluated the situation as follows:

> Increasing public concern about protection of environmental quality is matched by a corresponding expectation that higher standards will be set and greater attention given to achieving sustainable use of natural resources. Managing natural resources requires skill and expertise and maintenance of a data base. The data base is extremely important. New Zealand has lost many of the centralized data systems essential for identifying trends and essential for management systems. The data bases for water quality, soil condition, air quality, require scientific skills. These skills are having to be extended in local government as a result of central government shedding responsibility in water and soil management, and air pollution control. . . . Because the Act is dealing with effects, data bases are essential to identify whether in fact changes in natural systems are occurring and whether the changes are of any significance.[6]

A final set of expectations has to do with the nature of the relationship between regional and local governments. The clear intent of the *Resource Management Act*, as subsequently interpreted by the Planning Tribunal, is one for which regional councils are viewed as partners with, rather than superiors to, local councils. Regional and local councils are required to consult with each other, and local governments are obliged to consider whether their actions are "not inconsistent" with the policies and plans developed by regional councils. A ruling by the Planning Tribunal rejects the concept of a strictly hierarchical relationship while emphasizing the complementary roles of regions and districts under the act.[7] Regional governments are encouraged to develop a partnership with the local authorities in carrying out respective environmental management functions. As noted earlier in this chapter, there are a variety of reasons as to why these relationships would be less than positive. This draws attention to the character of relationships between regional and local government as an additional benchmark for evaluating the success of regional governments.

Capacity for environmental policy-making

Regional councils were consolidated from the prior, multiple regional entities and recast to fulfill integrative roles in environmental management. This would seem to suggest that regional councils in New Zealand were burdened by several handicaps. These include a legacy of prior relationships with local councils (and central government) that were not always positive, inheritance of staffs that were not necessarily well suited to the new tasks, and limited experience with the type of integrative environmental management that was expected. Cynics might argue that regional councils were handicapped from the outset.

Many central government policy-makers saw the re-casting of regional councils as a pragmatic way of moving quickly to establishing a strong regional government role. The new regional entities were a logical consolidation of what was in place, and the boundaries for the predecessor catchment boards made sense from a resource management perspective. Central government officials, however, recognized variation in funding and abilities among the regional councils. Indeed, the Ministry for the Environment recognized this by establishing a sliding scale of subsidies for regional council planning activities. From this perspective, the regional role was to be taken seriously and central government was doing what could be done, given limited resources, to shore up weaknesses.

Our assessment of the capacity of regional councils consists of responses about the commitment and capacity of governments for environmental and hazards management that we gathered as part of surveys in New Zealand. As with other aspects of the *Resource Management Act*, a caveat for this assessment is that it was undertaken early in the implementation of the legislation. Table 6.1 provides a comparative assessment of the perceived commitment and capacity of staff in district councils, regional councils, and unitary authorities. There are no noteworthy differences among the three forms of subnational government in terms of either perceived commitment to, or perceived capacity for, carrying out environmental and hazard-management functions. The mean scores for each category of jurisdiction are higher than one might expect given the extent of change in local government structures. The ratings respondents gave to commitment for each type of government have an average score of over five on a seven point scale. Regional councils on average have slightly higher perceived capacity than district councils or unitary authorities. But these are not large differences.

The differences in the composition of staff from regional and district councils are more striking. Regional councils have on average seven times the number of environmental science staff as the corresponding staff of district councils.[8] In addition, staff of regional councils report greater familiarity with use of computer modeling of environmental phenomena and effects than counterparts in district councils. These differences reflect previous roles of regional councils in carrying out such tasks such as water resource planning that involved environmental

Table 6.1 Commitment and capacity of local governments (New Zealand)

Dimension	Mean values for comparison of [a]		
	District Councils	*Regional Councils*	*Unitary Authorities*
Perceived commitment to hazard management	5.14 (1.25)	5.13 (.86)	5.25 (.87)
Perceived capacity for hazard management	4.27 (1.35)	4.72 (1.15)	4.50 (.43)

Source: Compiled by authors from survey of local governments in New Zealand
Notes:
[a] Cell entries are respective mean values for fifty-five district councils, twelve regional councils, and four unitary authorities. Standard deviations are reported in parentheses. Each item is scaled from 1 (low) to 7 (high). The differences in mean values for each dimension are not statistically distinguishable.

science and modeling. In contrast, district councils had experience in broader community planning and with social programs.

While our measures of staff capacity can hardly be considered definitive, they provide some evidence that regional councils have at least the minimal capabilities for the role assigned to them. As is also the case for district and unitary councils, the policy and planning roles assigned to regional councils present Herculean tasks largely because of the shift from activities-based to effects-based planning. The latter puts a premium on ability to predict and mitigate adverse environmental effects of development. As noted from the comments cited earlier of the Parliamentary Commissioner for the Environment, Helen Hughes, the early evidence suggests that these systems are not in place and will be difficult to develop under a devolved system of environmental management.

THE PERFORMANCE OF REGIONAL GOVERNMENTS IN NEW ZEALAND

While facing difficult challenges, New Zealand's regional councils do not seem to be hamstrung from the outset. A foundation for success is provided by a combination of the legislative provisions that grant explicit powers to regional governments, central government funding (albeit modest in amount) to support regional planning, staff expertise described in the preceding section, and independent political and financial bases. We note three benchmarks for evaluating the performance of regional councils: (1) the quality of regional policy statements; (2) the extent to which they are performing facilitative roles for local planning; and (3) the character of relationships with local councils. What follows is an assessment of performance of regional councils with respect to these benchmarks, based on the first two years of experience under the *Resource Management Act*.

Mixed performance in developing regional policies

One indicator of the performance of regional councils under the new regime is the experience in producing required policy plans. Each of the regional councils and unitary authorities produced the required draft policy statements for public notification within a tight two-year time constraint, although four of the draft statements were submitted several months past the initial deadline. The reasons for these delays varied and included funding and staffing problems, concerns about whether the requirements for cost–benefit analysis of policies had been met, lack of experience in plan preparation, and the disestablishment of one regional council and consequential creation of three unitary authorities in its place.

This is a strong record in meeting procedural deadlines for policy statements, especially because most regional councils were also absorbed in preparing compulsory regional coastal plans. Moreover, there is no tradition in New Zealand of central government taking punitive action with respect to local governments who fail to prepare plans or policies on time. Given other commitments, limited funding, and lack of a stick to coerce compliance, it is surprising that the regional governments have been this responsive. One explanation relates to the instability of regional entities in New Zealand, which fostered a need to demonstrate their contributions to resource management. To quote one central government official's assessment of the situation, senior managers in regional councils were highly motivated "to get runs on the board."

Although the regional councils generally met the tight deadlines for producing initial policy statements, the quality of those statements was uneven at best and generally viewed by central government officials as being disappointing. The key components of these criticisms are aptly characterized by the Deputy Secretary of the Ministry for Environment (MfE) in stating:

> To date, and from MfE's perspective, the results evident in notified regional policy statements have been generally disappointing. Many, but not all, policy statements do not seem to us to yet be able to fulfill the functions intended for them. Problems that concern us are:
> 1. There is too much very general material in many statements. . . . Conversely, some methods are too prescriptive and virtually de facto rules;
> 2. A number of natural resource users (at least the ones with whom MfE consults) consider that the generality of many statements provides them with uncertainty . . .
> 3. The vital ingredient of integration – among policies and resources – is not sufficiently evident . . .
> 4. Some statements seem more like work programs; they say little about policy, except that it is still to come . . . [9]

Some councils have not paid sufficient attention to whether proposed objectives and policies are within their powers. In many instances, there appears to be poor integration of relevant planning issues at a policy level. The statements do not usually indicate in what way integrated management will take place on a region-wide basis, and thus include both regional and district planning considerations. Given that the main purpose of the regional policy statements is to provide a mechanism for integrated resource management, this feature is disappointing. Staff of some regional councils question the extent to which such integration will ever occur, given what they see as basic constraints on regional council authority and limitations of the *Resource Management Act*.[10] The first generation policy statements produced by the regional councils are very similar in approach, but are likely to be more differentiated by the time the submission, hearing and appeal processes are completed.

In order to quantify aspects of quality, we rated the regional policy statements with respect to their goals (the extent to which goals for hazard management were specified) and policies (the extent to which policies were specified for hazards management).[11] The scores for these dimensions of the hazard management components of the regional policy statements show the qualitative criticisms cited by others. The statements score relatively highly with respect to delineation of goals (averaging seventy-one out of a maximum score of 100), but are much weaker in presenting policy direction (averaging eleven out of a maximum score of 100). Thus, while goals are generally well articulated, the translation of those goals into policies is much weaker.

The higher scores on the goals dimension reflect the fact that a central purpose of regional policy statements is to provide a statement of objectives in identifying relevant resource issues in each region. The weaker score on policies may be accounted for by the fact that the optional regional plans are viewed by some staff of some councils as a vehicle for specifying appropriate actions. As noted in the preceding commentary, staff of the Ministry for the Environment suggest that this reflects confusion among regional councils over the differences between policy statements (contained in regional plans) and policy implementation (contained in optional regional plans).

We are able to examine variation in the policy statements of regional councils by combining data from our survey of regional councils with our ratings of the quality of their policy statements. Our explanatory factors are a mix of measures of variation in the planning situation (staff professionalism and the year the policy was prepared), development situation (amount of hazard-free land and amount of development in hazardous areas), and degree of risk (prior losses, risk from flooding, and risk from earthquake hazards). The statistical findings show that staff professionalism is an important factor in influencing the quality of policy statements.[12] In addition, councils with greater amounts of hazard-free land had lower-quality regional policy statements, reflecting what we depict as less pressure to address the problems posed by development in hazardous areas. As the amount of development in hazardous areas increases, the range of planning tools that can be applied is restricted, thereby creating less opportunity to develop broad-ranging policies. We found that to be true here, with regional councils proposing fewer hazard mitigation policies where more development was located in hazardous areas. These results also show that prior experience with losses from floods or earthquakes and greater risk of future losses stimulate planners to develop goals and policies for hazard mitigation.

Successes in performing facilitating roles

At the same time that they come to terms with their responsibilities for regional policy development, regional councils are expected to demonstrate their abilities as planning facilitators and information brokers in helping local governments. This role is essential for realizing the full potential of devolution in helping to create innovative policies and practices which reflect local circumstances. As such, the extent to which regional councils are fulfilling these facilitative roles is an important benchmark for intergovernmental environmental management.

To varying extents, regional councils have acted as a resource for local councils. The evidence is that the facilitative role is more variable than envisioned, but in general regional councils are being called upon for assistance. Our survey of local councils shows that local council staff turn more often to regional councils for information and assistance in developing policy and plans than they turn to other local councils, or to the central government's Ministry for the Environment or Department of Conservation. The extent to which local councils rely on the different sources of information for policy guidance ("reliance") and the extent to which staff in local councils report the guidance is relevant to their needs ("relevance") are reported in Table 6.2. Reliance on regional councils is on average about 50 percent greater than the other sources of information. Staff from local councils report relevance of regional councils to their situation being greater than other sources of information by similar margins.

Our efforts to statistically model the variation in local council reliance on regional councils for assistance demonstrate a set of factors that provide a common bond and another set of factors that divide the two levels of government, at least as they apply to the management of natural hazards.[13] The common bond is fostered by shared risk, experience with losses, and demand for development in hazardous areas. As each of these takes on higher values, there is increased reliance by local councils on regional councils. To a lesser extent, as each of these increase local councils also see regional councils as more relevant. This makes sense since local and regional councils are likely to face similar magnitudes of threats from hazards and development pressures. Moreover, as noted earlier, regional councils tend to have greater scientific expertise for assessing environmental effects.

The divergence in the reliance of local councils on regional councils is fostered by a set of political and organizational factors. Our modeling shows that with increased constituency demands to address hazards problems and enhanced levels of the capacity of local councils, the latter become less reliant on regional

Table 6.2 Intergovernmental sources of advice and assistance (New Zealand)

District Council advice or assistance provided by:	*Mean values for intergovernmental asssistance*	
	Reliance[a]	*Relevance*[b]
Regional Councils	6.02 (0.98)	5.74 (1.01)
Other District Councils	3.89 (1.84)	3.96 (1.69)
Ministry for the Environment	4.18 (1.89)	3.62 (1.45)
Department of Conservation	4.11 (1.69)	3.44 (1.70)

Source: Compiled by authors from survey of New Zealand local governments
Notes:
[a] Cell entries are mean values for the extent that district councils rely on other governmental sources for advice or assistance. The reliance index is scaled from 1 (low) to 7 (extensive). Standard deviations are reported in parentheses.

[b] Cell entries are mean values for the extent that district councils find the advice or assistance of other governmental sources relevant. The relevance index is scaled from 1 (low) to 7 (extensive). Standard deviations are reported in parentheses.

councils. These factors help foster the commitment and ability to go it alone, and as a consequence local officials see regional councils as being largely irrelevant. These findings also suggest that regional councils are important sources of information when a local council is just beginning to come to grips with an issue, when the issue is of low salience, or when local councils have limited resources. As such, regional councils appear to be fulfilling an important backstop role in filling the void created by a lesser central government role.

Uneven relationships between regional and local councils

A final set of expectations has to do with the nature of the relationship between regional and local governments. The clear intent of the cooperative policy is that regional councils are viewed as partners with, rather than superiors to, local councils. The preceding section provided evidence that the desired consultation between regional and local councils is taking place to varying degrees. Also at issue is the character of relationships between regional and local governments.

As noted earlier in this chapter, there are a variety of reasons as to why these relationships would be less than positive. In some instances, there were difficult relationships between the predecessor regional authorities and local councils. One other complicating factor, which is specific to planning for natural hazards, is the assignment of dual functions for such planning under the *Resource Management Act*. The intent was explained by the Secretary for the Environment who commented:

> This dual responsibility was quite deliberate. It reflects the fact that some hazards are best managed at the territorial level and others at the regional

level. . . . With both levels of local government having functions under the Act, good communication between them is vital if the Act is to operate effectively.[14]

An attempt to clarify the boundary of responsibilities between both councils was made in an amendment to the *Resource Management Act* in 1993. This enables regional councils to identify in their regional policy statements the responsibilities of local councils for the management of natural hazards in all or part of a region. If no responsibility is identified in the statement, the regional council retains responsibility for the management of the natural hazard. This amendment did not eliminate key ambiguities to the satisfaction of some councils as reflected by the fact that the Canterbury Regional Council and Christchurch City Council have sought clarification from the High Court of a decision by the Planning Tribunal concerning the division of responsibilities between them as it relates to the hazard-planning provisions.

The ability to sort out this planning relationship serves as one indicator of the strength of the intergovernmental partnership. A review of twelve proposed regional policy statements by Hinton and Hutchings showed that two-thirds of them adopted an approach that could be considered a partnership under which regional councils focused on hazard threats of regional significance and local councils addressed more localized threats posed by natural hazards.[15] Two regional policy statements showed no division of responsibility so that primary responsibility for all hazards was retained by the regional councils. The remaining two regional councils acknowledged a division of responsibility, but did not provide guidance to their local councils on how this was to be achieved.

Other evidence about the intergovernmental relationships is provided by observations of these relationships, which in some instances, suggest emergence of a healthy set of intergovernmental relationships. For example, Wellington Regional Council and its constituent districts were able to build on positive working relationships to explore the potential of the new legislation. The working relationship between the Bay of Plenty Regional Council and the Tauranga District Council also illustrates cooperation that led to new approaches to environmental management. Faced with pressure for the intensification of development along its coast, planning staff in Tauranga District Council sought expert advice from consultants on how to delineate the coastal hazard zone and options for managing land uses in it. In so doing, they also sought the support of the Bay of Plenty Regional Council. Although the two began with different foci in addressing coastal hazards, workshops held among the two sets of parties and their consultants led to an interchange of information, and a broadening of the consideration of hazards for both jurisdictions.

However, such intergovernmental cooperation is by no means universal. Even within the Bay of Plenty region there is a counter example where another district council was much less willing to work with the regional council on coastal management problems. In addition, there is an acknowledgment by some regional councils, such as Otago and Canterbury, that their respective primary cities, Dunedin and Christchurch, have the wherewithal to proceed independently. Nowhere is the discord between regions and districts as pointed as it is in Auckland. Strong, well funded district councils have actively countered such regional initiatives as establishing urban growth limits. The regional policy statement has been criticized for going too far in terms of bounding territorial council activity and being overly prescriptive.

Perhaps the most telling indicator of the strength of the relationship is that a number of regional councils have survived efforts by constituent district councils to form unitary authorities – that is, entities with both regional and district council functions. Over the first five years of their existence, at least half of the regional councils experienced a challenge by one or more constituent councils for more autonomy.[16] Yet, these efforts were rarely successful. Some might see the desire of some district council officials to create unitary authorities as evidence of poor relationships with regional councils. Others might see the lack of success as an endorsement of the viability of regional councils. In either case, intergovernmental tensions are clearly evident.

CONCLUSIONS

This chapter has addressed the role of regional governments in resource and environmental management. The fact that regional entities often have precarious standing in intergovernmental frameworks is at odds with the rationality of seeking regional-level solutions to environmental problems. New Zealand's experiment in resource and environmental management provides an exceptional case, in comparison to other countries, for examining the roles of regional governments. As part of the reform of local government, policy-makers crafted regions with meaningful boundaries for resource and environmental management and provided them with independent political and financial bases. As part of the reform of resource and environmental management, policy-makers gave regional councils important roles in resource and environmental management. This was a serious set of reforms that serves as an exemplar of a systematic, rational approach to environmental management.

The primary issue of this chapter is the extent to which regional governments are fulfilling their promise. We have assessed that promise with respect to three sets of benchmarks: (1) their performance in preparing required policy statements; (2) the extent to which they are undertaking facilitative roles in providing technical assistance and a coordination role for local government planning; and (3) the character of the intergovernmental relationships between regional and district councils. Our data about the commitment to, and capacity for, regional councils to carry out aspects of the *Resource Management Act* suggest that they are not hamstrung from the outset. Their strengths are a relatively capable base of staff with scientific expertise and technical skills. Yet, they are constrained by a history of relationships with local councils that is not always positive and by inexperience in conducting the type of policy planning called for under the newer environmental legislation.

The performance of regional councils in New Zealand in producing policy statements is mixed. On the one hand, most regional councils met the tight deadline for producing draft policy statements for public review. On the other hand, the quality of the statements, in their draft form, is variable, but generally weak. This points to additional concerns about whether regional councils have the capacity to develop and maintain the environmental databases necessary to support sound policy-making and planning.

At the same time that they come to terms with their responsibilities for regional policy development, regional councils are expected to demonstrate their abilities as planning facilitators and information brokers for local environmental management. Here the record appears to be better, but still variable. The withdrawal of substantial central government guidance and assistance, has put the spotlight on the regions. Our survey of local councils shows that their staff turn more often to regional councils for information and assistance in developing policy and plans than they turn to other sources of government information. Our statistical modeling of regional and local relationships for natural hazards planning show that such reliance is fostered by a common bond of disaster experience and exposure. Yet with increased constituency demands to address hazards and enhanced local capacity, local councils become less reliant on regional councils.

The third, and perhaps most important, benchmark is the character of relationships among regional and district councils. The clear intent of the co-operative policy is that regional councils are viewed as partners with, rather than superiors to, local councils. There appears to be a serious effort to establish good working relationships, but in some instances these are hampered by the legacy of past conflict. As a consequence, we observe very positive relationships in some instances and strained relationships in other settings.

Taken together, these results show that the regional government role is being taken seriously in New Zealand. Some two years after being charged with important resource and environment management functions, regional councils are moving quickly to establish a positive record of accomplishment. These results are in some ways surprising, given the past precarious political standing of regional entities and a legacy of uneven relationships with local governments. Yet, the early record is mixed. Critics can point to a number of problems, and suggest that the regional role is not being adequately fulfilled. These include the restrained relationships that are evident for some regions, the poor quality of some draft policy statements, and the lack of sufficient regional-level monitoring of environmental effects. Yet, optimists can point to early successes, and suggest that regional governments have overcome major hurdles in establishing a viable role in environmental management.

Having considered the regional dimension of the cooperative approach to intergovernmental environmental management in this chapter, we can point to preconditions for an effective role for regions. More is involved than simply legislating a regional government role. There needs to be a strong commitment to the concept of a regional approach to environment management and to making that work. In the case of New Zealand, these consist of defining regions with respect to meaningful natural boundaries and vesting regional governments with important responsibilities. As important is making sure that regional entities have adequate political standing and capabilities to carry out their functions.

NOTES

1 This logic is developed in John Dryzek, *Rational Ecology* (Oxford: Basil Blackwell, 1987). For discussion in the New Zealand context see S.K. Wypych, "Wellington Regional Boundary Determination," Report to the Legislation and Regional Government Subcommittee (Wellington: Wellington Regional Council, 1988).

2 For an overview of this experience see: Michael Bradshaw, *Regions and Regionalism in the United States* (London: Macmillan, 1988).

3 This discussion about regional planning in Florida is based on: Florida Advisory Council on Intergovernmental Relations, "Substate Regional Governance," Council Discussion Paper (Tallahassee, FL: Advisory Council on Intergovernmental Relations, 5 November 1991).

4 For the prior history see: Neil Ericksen, *Creating Flood Disasters?* Water & Soil Miscellaneous Publication No. 77 (Wellington: National Water and Soil Conservation Authority, 1986), p. 159.

5 Helen Hughes, "Environmental Management and Regional Councils," Report of the Parliamentary Commissioner for the Environment, 26 September 1991 (Wellington: Parliamentary Commissioner for the Environment Office, 1991).

6 Helen Hughes, "Local Government Environmental Management With Particular Reference to Unitary Authorities," Address to Public Meeting, Invercargill, 29 July 1993 (Wellington: Parliamentary Commissioner for the Environment Office, 1993), p. 2.

7 "Decision No. A 89/94 in the matter of the *Resource Management Act 1991* and an application by the Canterbury Regional Council under Section 311 for declarations," Planning Judge D.F.G. Sheppard presiding (Christchurch: The Planning Tribunal, November 1994).

8 Calculated from responses to our survey of New Zealand governments. The mean numbers of environmental science staff are 8.6 for regional councils and 1.2 for district councils. For both categories of staff the variation in staffing levels is much greater across district councils than among regional councils.

9 Lindsay Gow, "Regional Policy Statements: State of Play," speech to the New Zealand Local Government Association Conference, 3 June 1994 (Wellington: Ministry for the Environment, 1994): 2. Also see reviews by: Suzanne Baird, "Overview on Practice Issues in Implementing the Resource Management Act," paper presented at the New Zealand Planning Institute Conference, 27–30 April 1994, Nelson (Wellington: Minister for the Environment, 1994); Jennifer Dixon, "Strategic Environmental Assessment in New Zealand," paper presented at the Fourteenth Annual Meeting of the International Association for Impact Assessment, 14–18 June 1994 (Hamilton, New Zealand: Centre for Resource and Environmental Studies, University of Waikato, 1994).

10 See, for example: John Hutchings, "Interauthority Integration of Approaches and

Issues: Regional Approaches," paper presented at the New Zealand Planning Institute Conference, 27–30 April 1994, Nelson (Stratford, New Zealand: Taranaki Regional Council, 1994).

11 Policy statements for twelve regional councils and two unitary authorities were rated. The rating procedure followed that employed by Philip Berke who assisted in the design of this portion of the research. The coding involved coding for the presence or absence of a series of potential hazard management goals and policies as discussed in the methodological appendix of this book. Also see: Philip Berke, "Evaluating Plan Quality: The Case for Sustainable Development in New Zealand," *Journal of Environmental Planning and Management* 37, no. 2 (1994): 153–167.

12 When controlling for the other factors noted here, the standardized regression coefficients for the variables discussed in this paragraph are as follows. For the regression explaining variation in goal quality (adjusted R2 of .93, F = 20.78 for eleven cases), the standardized coefficient for staff professionalism is .60 (p <.05), amount of hazard-free land –.47 (p <.1), and prior losses 1.05 (p <.01). For the regression explaining variation in quality of policies (adjusted R2 of .86, F= 9.68 for eleven cases), the standardized coefficient for staff professionalism is .55 (p <.05), amount of development in hazardous areas –.99 (p <.01), and risk from flooding .43 (p <.1).

13 This entailed a regression analysis using the rating of reliance on regional councils as the dependent variable and a set of explanatory variables with standardized coefficients as follows (adjusted R2 of .26, F= 2.87, n= 39 local councils): demand for development in hazardous areas .24 (n.s.), prior losses from disasters .26 (p <.05), constituency demands for addressing hazards –.38 (p <.05), risk of flooding .33 (p <.05), risk of earthquake .15 (n.s.), professionalism of planning staff –.2 (n.s.), and staff capacity –.19 (n.s.), where "n.s." indicates statistically non-significant at the .05 level or below.

14 Roger Blakeley, "Natural Hazards and the Resource Management Act," Opening Address to the Natural Hazards Management Workshop, Wellington, 8 November 1994 (Wellington: Ministry for the Environment, 1994).

15 Sheryl Hinton and John Hutchings, "Regional Councils Debate Responsibilities," *Planning Quarterly: Journal of the New Zealand Planning Institute* 115 (September 1994): 4–5.

16 See Hugh Fyson, "Breaking up the Regions?" *Terra Nova* 17 (June 1992): 7–8.

7

LOCAL PLANNING,
COMPLIANCE,
AND INNOVATION

A key aspect of intergovernmental environmental management is the development of plans by local governments. Each of the policy mandates we consider specifies a planning process. However, as described in Part I of this book, the nature of the process differs. Florida's growth management program is strong in prescribing the elements of local plans, while the environmental management legislation of New Zealand and the flood policy of New South Wales are less prescriptive with respect to plan content. Another key difference is the nature of higher-level government review of local plans. The key distinction is between the coercive approach of Florida and the cooperative approaches of New South Wales and New Zealand.

We shift focus in this chapter from the regional level of the preceding chapter to the local level of government. Of interest are how the local planning processes have played out over time and how the different intergovernmental approaches affect these processes. This involves considering different aspects of local compliance with policy mandates. One aspect is procedural compliance as measured by the extent to which local governments adhere to the prescribed planning processes. Other aspects are process considerations involving the extent to which the different policies have been successful in altering the local planning process so as to fulfill various aims. Of particular relevance in this regard is the nature of innovation in local planning. One of the promises of cooperative intergovernmental policies is that increased flexibility permits more local innovation than might be the case under a more prescribed, coercive policy where innovation originates at higher levels.

The issues addressed in this chapter concerning local governmental compliance with planning requirements, changes in planning processes, and the content of local plans get to the heart of intergovernmental environmental management. As noted above, planning requirements are cornerstones for the policies we consider. Of interest is how local governmental willingness to plan, attention to different planning considerations, and the quality of plans is affected by the two types of intergovernmental mandates. We get at this in this chapter by reviewing the differences in planning processes, assessing the extent of procedural compliance under the different mandates, characterizing changes in local planning approaches, and assessing differences in resultant plan content.

PLANNING PROCESSES IN THREE SETTINGS

Planning has developed as a discipline and profession to help decision-makers find mechanisms for managing development and balancing community interests when making investment decisions. Local government plans set forth objectives and strategies for managing development, and/or they specify rules for development. They have evolved as important tools that policy-makers use in guiding the development of cities, regions, or special districts. As discussed in this section, the nature of such plans is shaped by the requirements of the higher-level policies that either authorize or mandate local (or regional) planning.

Despite the differences in the nature of the mandates in Florida, New Zealand, and New South Wales, there are common elements of the planning processes under each. These include requirements or strong incentives for local planning, provisions for public participation and consultation in preparing plans, and provisions for review by higher-level authorities of resultant plans. The key differences in planning processes stem from the nature of the mandates and from distinctions in the types of plans that are required by each mandate.

Distinctions in types of plans

One of the more confusing aspects of planning processes is the difference in types of plans that are involved. These differences stem both from distinctions in policy requirements and from different planning traditions in each country. Three types of plans are involved: policy plans, functional plans, and regulatory plans. Because some of the planning processes involve plans that contain elements of more than one type of plan, even this categorization can create confusion.

The clearest distinction is between floodplain management plans prepared in New South Wales under the state's flood policy, and the types of plans produced in Florida and New Zealand. The floodplain management plans are functional plans that are prepared for the relatively narrow purpose of addressing flood hazards. As discussed in Chapter 4, the planning process entails development of a strong fact base about the flood hazard and consideration of a range of land-use and other solutions to the flood problem. Given their narrow focus and strong analytic component, these plans can be expected to be concrete guides to local decision-making. Their obvious weakness, however, is that in addressing a single purpose they ignore other aspects of community development that may affect, or be affected by, flooding.

The regional policy statements in New Zealand and both regional and local comprehensive plans in Florida can be described as policy plans. These are much broader in their orientation, and hazard management is but one of a number of objectives for environmental planning or growth management. A key question in the planning literature, not specifically addressed here, is whether comprehensive plans can serve as effective guides to local development management.[1]

The third type of plan is a regulatory plan that specifies the rules for development or other land use, and conditions for application of those rules. In the United States, these are usually contained in separate local zoning, subdivision, or other ordinances rather than being part of comprehensive (or policy) plans. Florida's planning mandate requires that such rules be consistent with policies specified in the comprehensive plan that is required of local governments under the growth management legislation. Under New Zealand's *Resource Management Act* the plans that are to be developed by regional and local councils incorporate regulatory components. Yet because the plans that local councils are required to prepare also contain policy components, the distinction between regulatory and policy plans in New Zealand is not that hard and fast.

Distinctions in process and review mechanisms

Public consultation and participation are increasingly important elements of the practice of planning. This trend recognizes that planning is a highly political activity and that it is important for planners to identify the needs and aspirations of various groups within communities in order to produce plans that are capable of being successfully implemented. Participatory planning is an important development in planning style where planners and communities work together to achieve common goals.

The three planning processes differ in the specifics of participation and consultation. Each involves requirements for consultation and participation, a process of formal notification of draft plans for public review, and provisions for submission of comments about draft plans and related hearings. New Zealand's legislation is the most prescriptive of the three policies with respect to these requirements. The legislation sets out the groups to be consulted in preparing plans and provides standing to anyone or any organization that wants to comment on draft plans, through the submission process, while also prescribing the timeline for accomplishing this. The growth management legislation in Florida is less prescriptive with respect to who is involved, but specifically requires citizen involvement in the planning process and sets forth timelines for participation. The legislation also gives citizens who believe participation requirements have not been met standing to challenge plans through state administrative hearing processes. The planning guidelines for the flood policy in New South Wales encourage community participation and facilitate it through creation of local floodplain management committees. In addition, the state's environmental planning legislation, under which floodplain management plans gain legal standing, contains participatory requirements and legal mechanisms that are generally similar to those of Florida's growth management provisions.

The reality of preparing plans under each of these processes is that there are many participants of whom planners represent one professional group. In Florida, local planners create development policies for guiding decisions about growth that are to be consistent with other policies. The latter include plans prepared by the relevant regional councils, resource plans prepared by water management districts, and beach and shoreline preservation plans. In New South Wales, committees of local government politicians, professional staff, and community representatives work together in preparing floodplain management plans. Usually, an engineer from the state's Department of Public Works or Water Resources Department will attend these meetings to provide technical advice, but the contribution from state government is relatively limited.[2] In New Zealand, planners largely take responsibility for preparing plans at the local level. As discussed in the preceding chapter, the situation is somewhat different in regional councils where scientists and engineers dominate staffing.

Also relevant are differences in review mechanisms by state (New South Wales and Florida) and central government (New Zealand) agencies. In Florida, local plans must be approved by the state's Department of Community Affairs as being in compliance with relevant requirements as reflected in a long checklist. If local plans fail to meet state requirements, the state can draw on a range of potential sanctions (discussed in Chapter 2) or negotiate agreements for bringing the local plans into compliance. Under New Zealand's environmental management legislation, final decisions about the content of local plans are much more in the hands of local and regional governments. Although the Ministry for the Environment can provide comments about local and regional plans (as can any other agency or individual), the Planning Tribunal has ultimate authority to hear appeals arising from disputes by parties involved in the process. The Tribunal does not review whole plans. Instead it can only consider the issues brought before it by those making submissions, and the range of remedies is much more restricted than in Florida. State authorities in New South Wales do not undertake a formal review of floodplain management plans that local governments create under the state's flood policy. However, as a practical matter, such review is almost always involved in order for local governments to obtain state funding to implement the plans.

Distinctions in substance

A major distinction with respect to the planning processes we consider is the focus of the planning. There is an important distinction between planning for different uses of land or activities on the one hand, and planning to achieve a desired set of effects or environmental outcomes on the other hand. Activities or use-oriented planning is the traditional approach of prohibiting certain actions (e.g., development within a specified coastal zone) and specifying rules that regulate the conditions for obtaining permits to undertake other actions (e.g., floor height requirements). Effects-oriented or outcome planning is a more open-ended search for means for achieving desirable outcomes or minimizing undesirable effects. Aspects of this are built into American environmental impact assessment requirements as components of national and state legislation requiring assessment of the environmental consequences of certain types of decisions or development.

As discussed in Chapter 3, the effects-based planning concept is taken much further in New Zealand with the integration of statutory planning and environmental impact assessment under the *Resource Management Act*.[3] The legislation signals a major shift in planning philosophy and procedure away from old style zoning and control of activities toward assessing the consequences of different actions. The flood policy of New South Wales seeks a broader assessment of the effects of development within floodplains and delineation of tradeoffs that are involved, than was the case under the prior, more prescriptive flood land-use policy. As noted earlier, however, this entails a much more focused set of considerations relating to flooding than that being attempted in New Zealand.

Some might argue that this distinction in planning foci is not that sharp, since well-trained planners under either system are taught to think about the effects of different plan provisions. However, the nature of the planning mandate provides constraints on practice. Here, we can expect to see distinctions between planning outcomes under prescriptive planning mandates, like Florida's growth management program, and those found under the less prescriptive planning mandates of New Zealand and New South Wales. In particular, we expect to find more locally formulated, innovative approaches to addressing environmental management under the cooperative, less prescriptive regimes than under the coercive, more prescriptive regime found in Florida.

PROCEDURAL COMPLIANCE

Procedural compliance is the extent to which local governments comply with process prescriptions in preparing required plans, such as meeting the designated deadlines, and meet other process-related requirements. The growth management legislation of Florida required local plans, and when local governments failed to meet deadlines for plan completion, state officials applied sanctions. The merits flood policy of New South Wales made floodplain management plans voluntary, but conditioned future flood-control funding and waivers of local liability on completion of the plans. New Zealand's environmental management legislation requires plans, and sets deadlines for their completion. But it only provides weak mechanisms for enforcing the deadlines.

The policies provide different foundations for the monitoring that is undertaken by higher-level governments of the procedural compliance by lower-level governments. In specifying subjects that plans must address and detailing consistency requirements, the Florida policy establishes a strong basis for monitoring procedural compliance. Establishing whether or not a given jurisdiction complied with Florida's requirements is fairly clear cut. Determining what constitutes an acceptable policy document is a matter of checking for required provisions, local development regulations, and supporting documentation.

Determining procedural compliance in New Zealand and New South Wales is more complicated. Compliance can be assessed at the general level of whether a plan or policy has been produced, but it is less clear what constitutes an acceptable plan or policy. The lack of prescription about plan content makes what constitutes an acceptable plan more arguable, and formal review of plans differs from that undertaken in Florida. Apart from occasional state-wide reviews of the merits floodplain management program, there is no ongoing monitoring of local government participation in New South Wales. New Zealand's Ministry for the Environment and Department of Conservation have active roles in reviewing and commenting on regional and local policy statements and plans, but formal acceptance of plans is only required for regional coastal plans. The key decision for these agencies comes at the time of draft plan submissions, when they must decide what comments to submit as part of the public comment period. As noted later in this chapter, each agency had developed draft guidelines for such review at the time of our study.

Despite the difficulties of assessing full compliance, the procedural step of actually preparing a plan or a policy statement on time is an important test of the abilities of intergovernmental environmental mandates to bring about change in the way that local governments manage the environment. We use secondary data in Florida and New Zealand and our survey data in New South Wales to assess compliance with procedural provisions. One caveat to this assessment is that New Zealand's local governments are not required to have draft plans completed until October 1996 – five years after the original legislation was enacted.

As shown in Table 7.1, procedural compliance by local governments was much greater in Florida than in New South Wales – settings for which sufficient time has passed to gauge procedural compliance. Only 2 percent (seven out of 457) of the local governments in Florida failed to meet the 1989 deadline for plan preparation, and by 1994 all of these had prepared plans. In contrast with the relatively high degree of compliance attained in Florida, eight years after the 1985 merits policy was adopted in New South Wales, 38 percent of the local councils for which the policy is potentially relevant had not started the planning process or had taken minimal steps, and only 37 percent actually had a floodplain management plan in place. Although it is too early to assess eventual compliance in New Zealand, approximately one-quarter of the local councils had prepared draft plans by the end of 1994 and officials estimate over half of the seventy-three local councils will have draft plans by mid-1995. Although the degree of procedural compliance in New Zealand remains to be seen, these figures show substantial differences in procedural compliance between Florida's coercive mandate and New South Wales' cooperative policy. Understanding these differences in procedural outcomes requires consideration of the review mechanisms employed in each setting.

Strong compliance in Florida

Planning in Florida has involved a learning process for both the state and local governments. Many local plans failed to meet the strict state standards when they were originally submitted, and sanctions were imposed on some local

Table 7.1 Procedural compliance with local planning provisions

Florida's Growth Management Planning Mandate
(data as of 1 January 1994)

Stage of planning process	Local governments	
	Number	*Percentage*
Required plan submitted for state approval (Initial deadline 1 July 1989)	457	100
Plan submitted by deadline	450	98
Plan not submitted by deadline	7	2
Plans in full compliance upon submission	187	41
Plans brought into compliance by negotiation	246	54
Plans not in compliance as of 1 January 1994	24	5
Required plan not submitted for state approval	0	0

New South Wales' flood policy
(data as of 30 September 1993)

Stage of planning process	Local councils	
	Number	*Percentage*
Recommended process being pursued (No deadline specified)	78	62
Flood studies undertaken	17	13
Floodplain and issue studies undertaken	15	12
Completed studies and plans	46	37
Recommended process not being pursued	48	38

New Zealand's Resource Management Act
(data as of 1 December 1994)

Stage of planning process	District councils	
	Number	*Percentage*
Required plan submitted for review (Initial deadline 1 October 1996)	17	23

Sources: Florida: Department of Community Affairs, "Tabulation of Compliance with Planning Requirements," Unpublished memo (Tallahassee: Department of Community Affairs, 1994)

New South Wales: Compiled by authors from survey of local governments

New Zealand: Ministry for the Environment, personal communication, January 1995 governments because of that. In most cases, however, state and local governments negotiated their differences so that few localities were actually forced to comply completely with state dictates. In the years from 1989 to 1993, 184 cases involved findings by the Department of Community Affairs of noncompliance with state planning standards, and four cases that originated by other petitioners were brought before the Florida Division of Administrative Hearings. Of these, 180 were settled by negotiation, and only four went to the stage of remedial action ordered by the hearing officer.[4] Typically, negotiated settlements involved a stipulated settlement agreement in which local governments were required to comply with remedial actions agreed to by the Department of Community Affairs and other intervenors.

Interviews in Florida by planning professor Judith Innes revealed that while local governments reached accommodations with the state, in many cases they felt coerced into an agreement favorable to the point of view of staff of the Department of Community Affairs (DCA). She writes:

> Local officials are under pressure to settle to avoid losing substantial state funding. They contend bitterly that DCA "gets what it wants." DCA staff agree. They see the negotiations as an efficient procedure for showing the localities that they "mean business." They do not regard the meetings as a way to give localities a greater voice.[5]

Staff of the Department of Community Affairs were serious about achieving strict compliance with deadlines and plan content requirements. Consider these actions taken with respect to hazard mitigation requirements. In 1991, the agency rejected the city of Jacksonville/Duvall County's comprehensive plan, in part, because it failed to meet state standards for floodplain management and protection of coastal resources. Another community adopted a plan that would have overloaded evacuation routes in the event of a hurricane emergency; in response, the state rejected the plan and recommended the locality reduce proposed development densities and intensities.[6] The state has also held local governments to a strict definition of hazardous areas. When the city of Lynn Haven sought to amend its plan to allow development in a high hazard area by relaxing its definition of areas at risk from hurricane hazards, the state rejected the amendment.

Yet, state officials in Florida were sensitive to the potential backlash that was festering by their taking a hard line in seeking local compliance. As noted in Chapter 2, state agencies offered a series of financial incentives to both build the capacity of local governments to plan for and manage development, and to lower costs to comply with state requirements. These financial inducements also had the effect of reducing local government resistance to the state dictates.

More limited compliance in New South Wales

The more limited compliance of local governments in New South Wales in developing floodplain management plans stems from a variety of sources.

Perceptions of council staff that they do not have a flood problem accounts for some of the foot dragging.[7] Of the thirty-nine local councils where staff thought they had a small or no flood threat, 77 percent had yet to initiate planning. In contrast, only 21 percent of the eighty-seven local councils where staff saw the flood risk as moderate to very severe had not started or taken minimal steps. Even so, just 44 percent of those with a moderate to very severe perceived risk had completed the planning process.

To get at the broader set of factors explaining variation in procedural compliance, we developed a statistical model that contrasted those councils in New South Wales that had not complied with the recommended process with those that participated in or completed the process. Our modeling suggests that several factors contribute to procedural compliance.[8] As expected, the perceived risk of flooding was a noteworthy factor in distinguishing the two groups. Other noteworthy factors were the commitment of elected officials to hazard programs, the population of the jurisdiction, and rate of population growth. The importance of commitment from elected officials is clear. Official endorsement is required for undertaking such planning efforts, and necessary for freeing up governmental resources to undertake the planning. The population of the jurisdiction serves as a proxy for resource capacity, for which those jurisdictions with greater resources are more likely to undertake the recommended process. (Fifty percent of councils with populations of more than 20,000 had completed the process by 1994, while only 27 percent of councils with populations of less than 20,000 had completed the process.) The noteworthy influence of growth suggests that some jurisdictions are attending to floodplain management in conjunction with addressing a range of growth issues.[9]

The approach that governmental agencies in New South Wales adopt in dealing with local governments also proved to be a noteworthy factor in our modeling of procedural compliance. To get at this, following the discussion in Chapter 5 of the implementation style of agencies, we included in our modeling the implementation style of the relevant state agency regional office. Those local councils in areas where state agencies adopted facilitative styles were more likely, controlling for other factors, to have participated in the recommended planning process than those councils in areas where state agencies adopted a more formal, legalistic style.[10] It seems that coercive agency dealings within a cooperative policy regime are likely to backfire. Instead, facilitative approaches are necessary to build local governmental willingness to participate. This is an important finding since it suggests that coercion is not the only route to improved procedural compliance.

Local governmental immunity also seems to be relevant as a carrot in inducing participation. Of those councils participating in the process, 81 percent report that the treatment of legal liability for floodplain related land-use decisions plays a moderate or major role in flood-hazard management. Only 32 percent of those not participating in the process report similar roles in their decisions. The liability amendments were cited by 73 percent of those participating in the process as having been moderately or very important in stimulating flood-plain management efforts.

Uncertain compliance in New Zealand

In the preceding chapter we discussed the adherence of regional councils in New Zealand to deadlines for submitting draft policy statements and the variable, but generally low, quality of those statements. The compliance of New Zealand's seventy-three local councils with requirements of the *Resource Management Act* for preparation of district plans is uncertain, since they have until October 1996 to complete them. The date of plan review for each council is determined by the date that their district scheme plan fell due under the old regime. Many local councils appear to be on track for submitting their new district plans by the relevant deadline. However, given track records under the previous regime, it is expected that a substantial number will miss their deadlines. Because there are few sanctions to induce local compliance, the ability to bring recalcitrant councils into procedural compliance will rest heavily on the willingness of central government agencies to intervene or use of ministerial powers for the same purposes. Given the potential for Planning Tribunal involvement in reviewing aspects of plans, and related appeals, it is likely to be ten years before the transition from old to new is completed.

One factor that restrains timely compliance by local governments with procedural requirements is the high cost of carrying out the effects-based approach to planning. This is exacerbated by the devolution of responsibilities to regional and local governments, which places even more burden for planning and services on them. The Minister for the Environment acknowledged in a speech to a New Zealand Planning Institute Conference that considerable costs have been transferred from central government to local government as a consequence of government reforms.[11] Similarly, the New Zealand Business Roundtable also recognized the costs involved in implementing the legislative provisions in a report about implementation of the legislation. The author of the report, Alan Dormer, noted that "the costs of transition are high, and the sooner doubts as to the efficacy and legality of many documents' techniques are resolved, the quicker those costs will be behind us and the promised benefits of the Act achieved."[12]

Unlike the Department of Community Affairs in Florida, central government agencies in New Zealand are not undertaking a stance of coercing local government compliance with prescribed planning provisions. The difference between the approach in Florida and the broader compliance assessment in New Zealand is evident from quoting New Zealand documents concerning review criteria:

> [Department Of Conservation]: The Department has two tests for regional coastal plans: (1) legal tests, and (2) purposive, or general analytical tests. . . . [Legal tests]: The regional coastal plan must be "not inconsistent with" the New Zealand Coastal Policy Statement. . . . How the NZCPS is implemented is at the discretion of the regional council. . . . [Purposive

135

assessment considerations]: that the issues, objectives, policies, and methods are appropriate; environmental objectives must be clear and measurable; the plan provides certainty; that the plan is reasonable.[13]

[Ministry for the Environment]: Our monitoring function cannot be separated from our advice, consultation, statutory and non-statutory advocacy. . . . We are not into quality control, nor telling communities what's good for them – beyond where they are contrary to the purpose of RMA. . . . Submissions/objections will be few in number but significant. We need to prove ourselves by our track record. Only then will early consultation and other means of communication be seen as credible. Council's political problems can't be ours unless it significantly affects RMA outcomes, i.e., don't play politics.[14]

One concern of local governments is that central government agencies seem torn between a consultative position and a legalistic position in commenting about draft plans submitted by local or regional governments. The experience with regional policy statements (addressed in the previous chapter) was that once the draft statements were submitted, the tenor of involvement by the Ministry for the Environment shifted somewhat from a consultative approach to a more formal and legal approach. This increased legalism created negative reactions among some regional councils. However, this approach is consistent with the nature of the statutory planning process in New Zealand in that positions often become adversarial once formal legal processes commence. Ministry officials made it clear that they would pursue issues which were not resolved prior to plan submission, through the courts if necessary. Nonetheless, the basic approach in New Zealand – particularly when contrasted with Florida – is one of limited higher-level government interference in the preparation of local plans.

PROCESS CONSIDERATIONS

More was intended with specification of planning processes than just producing a plan or policy statement. Each of the mandates sought to change the character of planning and decision-making about environmental and/or hazards management. Florida's legislation did this in a highly prescriptive manner, detailing the sections to be included in local plans and stipulating a rigid review process to assure that those sections were included. New Zealand's *Resource Management Act* set forth a planning process and the types of issues to be considered, but was less prescriptive and provided more discretion in the format and content of regional policy statements or local district plans. New South Wales' merits-based flood policy also set forth a recommended planning process and specified relevant considerations for that process, but was the least prescriptive in the content and form of resultant plans.

Of particular relevance is the extent to which the character of local planning has been altered under cooperative policies. In New Zealand, and to a lesser extent in New South Wales, changes that were sought included: enhancing participatory processes in planning; drawing attention to environmental considerations in plan preparation (and to extreme floods in New South Wales); and encouraging the use of benefit–cost and other techniques for evaluating plan provisions. Data about local planning processes that we collected as part of our local government surveys in New South Wales and New Zealand enable us to address resultant changes in planning processes.

Different aspects of changes in planning processes for local councils in New South Wales and New Zealand are shown in Table 7.2. For each process consideration, the percentage of councils that report increases in the item after introduction of the new policy and the percentage of respondents who thought the new policy was important in bringing about the change are shown. We restrict attention to those local councils in New South Wales that participated in the floodplain-management planning process, since we are interested in gauging the effect that following the process has on local planning.[15] For New Zealand, we consider all district councils because it is too early to differentiate full participants from non-participants. Once again a caveat applies that the New Zealand data were collected early in the transitional phase of the new legislation when most councils were just beginning to develop new plans.

The results show that the policies have brought about changes in the character of planning processes, but those changes are not uniform. The most noteworthy changes are the increases in the extent to which environmental considerations are part of planning processes. Respondents correctly perceive both sets of legislation as placing greater emphasis on environmental considerations than was the case under the prior legislation. For local governments in New South Wales or in New Zealand, there is also strong recognition of the importance of public participation in policy formation. However, the survey results show less evidence of increased participation in local planning processes. This probably reflects the fact that public participation was already an element of planning processes in both settings prior to the new mandates.

The remaining entries in Table 7.2 consider more technical aspects of planning processes. The flood policy in New South Wales placed emphasis on the need to plan for extreme floods, but it did not prescribe a uniform standard. A majority of those councils following the planning process recognized this, and nearly two-thirds report increased attention to extreme floods. Conflicts between provisions of the *Resource Management Act* and the *Building Act* over the means for determining flood standards might account for the smaller percentage of New Zealand respondents reporting that the resource management legislation brought about increased consideration of extreme floods. A majority of both sets of respondents recognized the increased emphasis on use of benefit–cost analyses as decision-making tools for plan preparation, but lower percentages of respondents report increased use of the technique. This may partly reflect the fact that council staff, by and large, are not well versed in applying this method or impact assessment methods generally.

137

Table 7.2 Process considerations in local planning

Process consideration and percentage reporting[a]	New South Wales Local councils[b]	New Zealand District councils[c]
Consideration of environmental issues in planning		
Increased consideration	70	67
Policy important for change	61	80
Public participation in the planning process		
Increased participation	53	29
Policy important for change	77	62
Consideration of extreme floods in planning		
Increased consideration	67	31
Policy important for change	56	61
Benefit–cost analysis use in planning assessments		
Increased use	42	23
Policy important for change	61	60

Sources: Compiled by authors from surveys of local governments

Notes:

[a] Cell entries are the percentage of respondents from local councils indicating a given response. Respondents were asked whether there was an increase, no change, or decrease for each procedural item under the mandated planning process. The percentage reporting decreases were less than four percent. Respondents were also asked how important the mandated planning process was in influencing the reported change. The "policy important for change" row is the percentage reporting "moderate" or "very" important responses.

[b] Local councils that have followed the planning process specified by the New South Wales' flood policy, based on responses from seventy-eight councils.

[c] District councils including unitary authorities, based on responses from fifty-nine councils.

INNOVATIONS IN PLANNING

One of the promises of cooperative intergovernmental planning processes is increased flexibility so as to permit greater innovation in local planning than might be the case under a more prescribed, coercive policy. One of the less explicit goals of the resource management legislation in New Zealand is to encourage local and regional councils to develop innovative approaches to resources and environmental problems. As noted in Chapter 3, a key thrust of reformers in New Zealand was to signal that central government wanted local governments to do business differently, both in process and substance. The legislation signals a strong desire for regional and local governments to incorporate less-traditional policy instruments into their plans or policies, and to go about the planning process in less traditional ways. These are stark contrasts with the previous regime where the plan was seen as the principal, if not only, means for achieving the goal of controlling activities.

The situation is different in Florida, where the growth management policy is more prescriptive and restrictive. Although innovation in local planning is feasible, and even desirable, the nature of the state review of local plans provides a strong restraint on local innovation – a topic we explore below. Nonetheless, Florida is recognized as having a highly innovative approach to growth management that involves a number of concepts that had not previously been employed (e.g., compact development and the provision of infrastructure concurrently with the approval of new development). The difference between Florida and New Zealand with respect to innovation is more a matter of which level of government is the focal point for innovation. In Florida, the state government was the innovator interest centered on securing compliance with the state-mandated innovations. In New Zealand, central government looks to the regional and local governments to be innovators.

We turn in this section to the evidence about a differential influence of coercive and cooperative mandates on plan content, which serves as a basis for making judgments about overall quality and degree of innovation (or unique-ness). The impression we get from secondary sources is one of uniformity to a fault of straitjacketing local governments under Florida's prescriptive planning process. In contrast, New Zealand's cooperative process seems to be resulting in much more variability in plan content while also producing some key inno-vations. Functional planning in New South Wales presents a different situation, which we also consider in this section.

Straitjacketing in Florida

Our interviews with planning officials in Florida and review of evaluations of local plans evidence a distinctive dynamic in the rush by local governments to comply with mandated deadlines for plan preparation. The emphasis seemed to be on "getting the required plan done" thereby introducing a rigidity in the planning process that we did not sense in New South Wales or in New Zealand. Local planners knew that a rigid checklist would be applied as part of the state review of plans, so the emphasis was in developing the required planning elements. According to David Powell, who served as executive director of the 1992 study committee that Florida's governor formed to review the state's growth management program:

> A recurring criticism of the local planning provisions is that the state role has resulted in a "cookbook approach to local planning;" many local plans have been criticized for being written to satisfy the DCA "checklist" and not to take the community to a specific future condition – a "destination" – that represents the shared vision of the community.[16]

Several planners in Florida who participated in interviews conducted as part of our study of American state planning mandates remarked about the rigidity of this process and offered pointed criticisms of the Department of Community Affairs, the lead agency charged with implementing the growth management program:

> [Interview 1]: The Department of Community Affairs should not act as a zoning board for the state. That type of local responsibility should be with local governments. Although the Growth Management Act is a good law, it is not administered correctly. . . . Local governments should be more in charge of their own destiny. . . . The Department of Community Affairs claims that the locals have control, but they do not in practice.

> [Interview 2]: The process is too cumbersome. . . . The state wants to see long-range goals, but they also want short-term implementation. . . . The Department of Community Affairs wants to encourage innovation, but time restrictions and limited funds force local governments to rush through the process just to get done with it.[17]

These comments suggest that the greater procedural compliance in Florida may come at a price of straitjacketing local governments.

Plan innovation in New Zealand

The evidence from regional policy statements and local plans in preparation suggests more limited innovation in policy or plan provisions to date than hoped for by those who crafted the legislation. The early response was characterized by a headline in New Zealand's national business publication as "feet-dragging as councils feel their way through the RMA."[18] As depicted in that article, some councils were waiting to see what other councils produced so that they could learn from the experience of others. Other councils appear to have clung to the view that their existing plans meet the demands of the new legislation. Shifting from a style of planning that was heavily oriented towards development control to one that is effects-based is never going to occur quickly, especially without significant central government assistance. The Minister for Environment seemed to recognize this problem when commenting about the development of new plans and hopes for future innovative provisions:

> While the pains of the transition from the old to the new have to be recognised, the signs are there that the Act's potential can be realised. While it is perhaps understandable that early decisions under the Act will take a fairly conservative line, this should not mean that opportunities for innovation and creativity are avoided.[19]

One difficulty that central government officials faced was determining the type of guidelines for plan preparation to provide to local councils. The experience with model codes under the prior planning regime resulted in the production of very similar district plans. Given this experience and the difficulties of producing appropriate and relevant documentation, there has been a reluctance by the Ministry for the Environment to produce how-to-do-it models of policies and plans. Instead, the agency has produced a series of guidelines discussing the philosophy of the legislation and related planning processes while encouraging flexible and appropriate responses by local councils. As stated in the introduction to the agency's guideline for district plans: "The Guideline is not designed to give territorial authorities [local councils] the blueprint for their plans; it provides the starting point for territorial authorities to develop their individual responses to statutory requirements."[20] Given the lack of resources for many councils and the pressures on staff to develop plans, some practitioners and developers argue that this approach has been less than helpful and "too little, too late."

One of the real constraints on innovation is the uncertainty among planners about how to operationalize effects-based planning. Many planners have been waiting for rulings to emerge from the Planning Tribunal on particularly weighty issues, such as the meaning of sustainable management. These have been slow in coming. One consequence of the uncertainty seems to be that councils are relying on the Planning Tribunal to produce decisions to clarify and guide planning practice. This reaction is not unexpected, as the Planning Tribunal has been influential in determining planning issues for the last forty years in New Zealand. In addition, the prospect of lengthy and costly arguments before the Planning Tribunal over plan provisions can deter the adoption of radically different approaches to planning and environmental management. This reliance on the Planning Tribunal has the perverse effect of fostering legal formalism in the planning process, which is the opposite of the type of innovation and creativity that the legislation seeks.[21]

The planning approaches of the emerging district plans fall into three main categories.[22] One category is comprised of plans that still rely on zoning, but specify assessment criteria and standards for evaluating proposed development. These are similar to the traditional planning approach. A second category is comprised of plans that operate by exception. They allow activities to take place as long as they are not part of a list of exceptions. The third category are the more innovative plans that involve a threshold approach to determine the assessment criteria that apply to a given level of the effects of an activity.

Despite uncertainties about how to proceed, some councils are emerging as leaders in developing innovative approaches that other councils can build on. For example, although it is still in development, the draft Christchurch City district plan has been heralded by Ministry for the Environment staff as providing a flexible, effects-based approach. Among other provisions, the plan establishes a series of thresholds of effects with which to determine whether or not a given proposed activity (or aspects of it) is allowed. Another plan that has been cited as an example of good practice is the Hauraki Gulf Islands district plan, prepared by Auckland City Council. Through use of base-maps and a series of information overlays, the plan shows the relationship between land capability, potential

141

environmental effects, and future sustainability. The plan delineates performance standards as a basis for determining appropriate activities, and as such avoids the old-style approach of listing permitted (or prohibited) activities. It also provides for flexible subdivision opportunities which are related to land capability, conservation and preservation, and management of environmental effects.

Other councils are developing innovative approaches as a consequence of their strategic planning activities, which also serve to frame plans prepared in response to the *Resource Management Act*.[23] (The role of the strategic plans in shaping sustainable development decisions was discussed in Chapter 3.) For example, officials of the Waitakere City Council, also located in metropolitan Auckland, have announced a vision of becoming an eco-city. The city has developed a "greenprint" for the future that identifies the key changes for Waitakere to reach this goal. This plan will help shape the specifics of the district plan, prepared under the *Resource Management Act*.

Functional planning in New South Wales

The merits flood policy in New South Wales provides an example of functional planning in asking local governments to address the specific problem of flooding. As discussed in Chapter 4, the recommended process entails technical studies of flood potential and detailed assessment of options for managing the hazard. Florida's growth management legislation requires local governments to consider hurricane and flooding as one aspect of the comprehensive planning process. While this might seem a minor consideration in overall planning in Florida, the United States federal government and State of Florida have invested heavily in developing factual data for use in local and regional planning for hurricanes. This includes development of flood-hazard maps, vulnerability assessments for coastal cities and counties, storm surge models that project areas that will be inundated, and other maps to aid evacuation planning.

Although we have not undertaken formal comparisons of plan quality between Florida and New South Wales, our perusal of plans indicates that those prepared in New South Wales have a much stronger factual basis and are far more detailed in their analysis and policy prescriptions than the hazard-related components of plans prepared as part of the Florida mandate. We attribute this difference to the nature of the two planning mandates. Because hazards are but one of a great number of policy issues addressed by the Florida comprehensive plans, these considerations get less detailed attention. Nonetheless, that attention appears to be greater than in the New Zealand plans where there is a more limited fact base for hazard-related planning. The functional planning for floods in New South Wales puts that item as the focal point, and entailed considerable expense in preparing detailed plans.

While functional planning would appear to be advantageous, there are clear drawbacks. One drawback is that functional plans cannot be used for every problem, since the result would be a plethora of plans, leading to considerable confusion. Relatedly, functional plans lose the integrative benefits of comprehensive planning in drawing attention to conflicts among solutions to different problems and in developing strategies that cross-cut problems. Finally, the experience in New South Wales demonstrates that when the interest of local officials in the problem being addressed is low, compliance with the single-purpose mandate is also low. Comprehensive-planning mandates, like that found in Florida, have the advantage of specifying a range of goals, at least some of which should attract local interest and spur local compliance.

CONCLUSIONS

This chapter sheds light on the strengths and limitations of coercive and cooperative policy approaches as they relate to local planning processes, compliance, and innovation. The coercive intergovernmental approach, as implemented in Florida, mixed with a certain degree of cooperation, has clear advantages in achieving compliance with procedural provisions. Facing the threat of sanctions, local governments in Florida produced plans in a timely manner. With the help of additional funds and negotiated compliance agreements, all of these plans were quickly brought into compliance with planning requirements. Local governments in New South Wales have much lower rates of compliance in preparing floodplain management plans, despite strong incentives to adhere to the process. Only those councils that are strongly committed to addressing the flood problem appear to be participating. It is too early to judge how local government compliance has played out in New Zealand, but the evidence to date suggests a pattern that looks more like that found in New South Wales than in Florida. The procedural compliance evidence shows that with a coercive mandate, state governments are better able to get local governments marching to the state's tune. This, of course, assumes that there is not substantial backlash to the coercive mandate that results in an unraveling of the whole planning process.

Both the New South Wales and New Zealand experiences indicate that cooperative mandates are capable of altering the character of local planning processes even if there are not strong sanctions for failing to meet planning requirements. In both settings, there was greater attention to environmental considerations after introducing new policies and particularly among those local governments that adhered to the planning processes. Among those complying with prescribed functional planning processes in New South Wales the quality of floodplain management plans appears to be higher than the quality of hazard-management components of local plans in Florida.

One of the key differences in responses under the two types of mandates is the degree of innovation. New Zealand's legislation explicitly sought such innovation, while Florida's growth management program imposed state-formulated innovation. Our qualitative evidence suggests that Florida's prescriptive and coercive provisions stifled local innovation by straitjacketing local planners. The New Zealand experience provides examples of innovation, but to date there appear to be fewer of them than hoped for or desired by central government staff. New Zealand commentators attribute this sluggishness to the newness of the policy and the uncertainty of what is being required. Problems of inadequate databases, inadequate funding, variable expertise, legal cautiousness, and the lack of suitable methods for practitioners to adapt have constrained the emergence of innovative approaches.

The findings of this chapter about procedural compliance and substantive performance in local planning suggest a dilemma. On the one hand, prescriptive and coercive mandates like Florida's growth management legislation foster a high degree of compliance that is not obtained under more cooperative approaches. The latter lack sanctions or other inducements to bring less committed and capable local governments into the process. On the other hand, the prescriptive and coercive mandates appear to limit opportunities for local innovation whereas the cooperative policies have that promise. Realizing that promise, of course, is not automatic as evidenced by the experience to date in New Zealand. This dilemma leaves a challenge that we return to in the final part of this book of figuring out how to motivate lagging jurisdictions that seem to require a coercive approach, while not straitjacketing leading jurisdictions that are capable of thriving under a cooperative approach.

NOTES

1 See: Raymond J. Burby and Linda C. Dalton, "Plans Can Matter! The Role of Land Use Plans and State Planning Mandates in Limiting the Development of Hazardous Areas," *Public Administration Review* 54, no. 3 (1993): 229–237.

2 Where flood management problems are more complex and cut across the boundaries of several local governments, there are different arrangements that involve a more significant contribution from state government. For example, in the case of the Hawkesbury River, the state's Department of Environment and Planning is funding the state's Department of Public Works to undertake local studies to advise on the flood management problem.

3 For further elaboration see: Jennifer E. Dixon, "The Integration of EIA and Planning in New Zealand: Changing Process and Practice," *Journal of Environmental Planning and Management* 36, no. 2 (1993): 239–251.

4 Andrée Jacques, "Florida's Local Government Comprehensive Plan Amendment Review Processes, Including an Assessment of the Changes to This Process from the 1993 Florida Legislative Session," unpublished paper (New Orleans: College of Urban and Public Affairs, University of New Orleans, 1994).

5 Judith E. Innes, "Group Processes and the Social Construction of Growth Management: Florida, Vermont, and New Jersey," *Journal of the American Planning Association* 58, no. 4 (Autumn, 1992): 440–453, at 444.

6 Florida Department of Community Affairs, *Technical Memo* 4, no. 3 (Summer 1989), p. 8.

7 The Gamma association between stage of procedural compliance with the prescribed planning process and perceived flood threat from a 100-year flood is .51.

8 The modeling consisted of a logistic regression that distinguished those councils that did not participate from those that participated in the process. We were able to obtain a good fit for our data (model Chi-square of 50.8; p-value <.01), involving

correct prediction of participation or not for 86 percent of the observations. The factors in our model included population growth rate, demand for development in hazardous areas, prior losses from flooding, population, perceived risk from flooding, commitment of elected officials to hazard programs, perceived influence of liability in floodplain management, and the implementation style of the relevant state agency regional office.

9 In a study of local regulation of development in earthquake-prone areas in the United States, Peter May and Thomas Birkland also found that population growth was an important factor in explaining differences in "leading" and "lagging" governmental attention to such development. See Peter J. May and Thomas A. Birkland, "Earthquake Risk Reduction: An Examination of Local Regulatory Efforts," *Environmental Management* 18, no. 6 (1994): 923–937.

10 This factor has a positive logit coefficient (one-tailed p-value <.05) thereby indicating that facilitative styles contribute to greater likelihood of participation.

11 Simon Upton, Minister for the Environment, "Address to the New Zealand Planning Institute," New Zealand Planning Institute Conference, Nelson, 21–24 April 1994 (Wellington: Minister for the Environment, 1994).

12 Alan Dormer, *The Resource Management Act 1991: The Transition and Business* (Wellington: New Zealand Business Roundtable, 1994), p. 63.

13 New Zealand Department of Conservation, "Draft Guide to Regional Coastal Plan Submission Preparation" (Wellington: Department of Conservation, 9 August 1994).

14 New Zealand Ministry for the Environment, "Manual for Monitoring RM Act," draft internal document (Wellington: Ministry for the Environment, n.d.).

15 Not surprisingly, the results for councils not participating in the merits process are substantially lower than the corresponding results for those that are participating. For those not participating: 2 percent reported increased public participation in planning; 32 percent reported increased consideration of environmental issues; 19 percent reported increased consideration of extreme floods; and 3 percent reported increased use of benefit–cost analysis.

16 David L. Powell, "Managing Florida's Growth: The Next Generation," *Florida State Law Review* 21, no. 2 (1993): 223–340, at 270.

17 Interviews conducted as part of research for: Raymond J. Burby and Peter J. May with Philip R. Berke, Linda C. Dalton, Steven P. French, and Edward J. Kaiser, *Making Governments Plan: State Experiments in Managing Land Use* (Baltimore: Johns Hopkins University Press, 1996).

18 *The National Business Review*, 8 July 1994; p. 26.

19 Simon Upton, Minister for the Environment, "Keynote Speech," to the Resource Management Law Association, 7 October 1994 (Wellington: Ministry for the Environment, 1994), p. 4.

20 Ministry for the Environment, "Guideline for District Plans" (Wellington: Ministry for Environment, n.d.): 5.

21 The legal influences on the planning process have been viewed as a constraint on planning practice in New Zealand under the prior planning regime. See: Ali P. Memon, "Shaking off a Colonial Legacy – Town and Country Planning in New Zealand, 1870s–1980s," *Planning Perspectives* 6 (1991): 19–32.

22 See Jennifer Dixon and Tom Fookes, "Environmental Assessment in New Zealand: Prospects and Realities," *Australian Journal of Environmental Management* 2, no. 2 (June 1995): 104–111.

23 This is discussed by Greg Vossler, "The Quest for the Environmental Grail," *Planning Quarterly* 115 (September 1994): 22–23.

8

SUSTAINABLE MANAGEMENT STRATEGIES

Local government adherence to planning processes is a means, but not an end in itself, to achieve the goals set by higher-level governments. One goal of each of the mandates we consider is to enhance the efforts of local governments to manage, but not necessarily restrict, land use and development in areas subject to hazardous events so as to reduce losses and promote sustainability. The origins of this objective stem from a history of unsatisfactory results from prior efforts to address natural hazards, particularly with respect to flooding.

For decades, structural controls – levees, dams, and other protection schemes – were promoted by national, state, and local governments as the solution to flood problems. Structural works held out the promise of reducing prospective losses from past developments at risk while also allowing more intensive future use of flood-prone areas. As policy-makers began to recognize the limitations of over-reliance on structural protection systems, attention shifted to other approaches for managing hazards from flooding and other natural events. Beginning in the late-1960s in the United States and a decade or so later in Australia and New Zealand, unified flood management policies began to incorporate mandates for non-structural provisions such as flood maps, land-use regulations, sub-division controls, and building codes. However, bringing about sustainable approaches to hazard management has proven to be elusive.

The stakes in identifying sustainable hazard management strategies are large. On the one hand, efforts to restrict development in hazard-prone areas have the potential for choking off the economic base of communities. On the other hand, past practices of heavy reliance on hazard-protection schemes raise the potential for longer-term catastrophic losses from events that exceed the scale of those for which the protective works were designed. A central question is the appropriate development of hazard-prone areas. The policies of New South Wales and New Zealand leave it largely, but not exclusively, to local governments to decide the answer to this question. Closely related to this question is the role that environmental values play in decision-making about development patterns. As noted in the last chapter, the policies in New South Wales and New Zealand have had some success in instilling greater attention to environmental values as part of planning processes.

146

The cooperative policies we study hold promise for enhancing hazard management, but they also pose challenges. One promise is that by shifting decision-making to the local level, approaches to dealing with hazards can be crafted that are politically more acceptable than those that can be obtained with the blunt prescriptions of higher-level coercive policies. Simply put, national or state land-use programs are not politically viable in the settings for which there is a tradition of home-rule and strong local property rights. A second promise is that by instilling a stronger environmental ethic in policy-making, local decision-makers will be more conscious of the economic and environmental tradeoffs they face. Appropriate decisions, in turn, have the potential for fostering environmental sustainability. This is a particularly strong theme of New Zealand's *Resource Management Act*.

The key challenge for these approaches is that of bringing about the desired changes, rather than simply falling back on old ways of doing business. The flexibility provided by a cooperative approach enables local governments to develop innovative responses to natural hazards appropriate to the local circumstances. While this provides the opportunity for decisions that are more politically acceptable than is possible through coercion, it has the potential for local councils to be captured by growth-oriented economic interests. A good land-use policy is one that guides decision-making in the right direction even when strong and persistent pressures to the contrary exist. Thus, overcoming such resistance poses the main challenge to local governments for successfully implementing the cooperative approach to environmental sustainability.

This chapter addresses the search for sustainable hazard-management strategies, focusing on floods in particular under the cooperative planning regimes found in New South Wales and in New Zealand. The discussion addresses the changes in hazards management by local governments that have occurred under cooperative intergovernmental regimes. Where appropriate, we make comparisons with experience under Florida's more coercive growth management regime. This discussion provides a practical assessment of the decisions that are being made by local governments. The empirical basis for our observations are data from our surveys of local governments and a set of case studies of local government decisions about flood-hazard management in New South Wales and New Zealand.

CHOICES IN HAZARD MANAGEMENT

Hazards researchers have characterized broad strategies for managing natural hazards.[1] The traditional strategy is to modify the nature of the event through structural means. This involves various engineering works aimed at reducing the frequency with which natural events affect urban and rural development. A second strategy is to modify the loss-susceptibility of communities. This entails a range of land use, development, and building controls aimed at reducing exposure of people and property to events.

There is a long history in Australia, New Zealand, the United States, and many other industrialized countries of attempting to use the first approach. Only in the past two decades has there been concerted effort by national or state governments to advocate land-use and building regulations to address hazards.[2] Several factors contributed to this interest, with developments in the United States preceding those of the other settings. One factor was recognition by various hazard experts of the limitations of structural solutions. A second factor was growing concern among policy-makers about the costs of protection and the prospects for catastrophic losses if an event overwhelmed the structural protection. A third factor was the environmental movement of the 1970s, which drew attention to the negative environmental consequences of structural protection that destroyed wetlands and re-channeled rivers.

In each of the countries, national or state hazard policies incorporated provisions that placed greater emphasis on non-structural approaches to hazard management, and in some instances there were efforts to reduce funding of structural works. Through requirements for participation in the National Flood Insurance Program, the United States government mandated that buildings be elevated to the level of the 100-year flood. Similar building standards were required by building codes in New South Wales, but in New Zealand they varied according to local circumstances. Environmentalists also began to call for land-use approaches to deal with flood hazards, so that the rich natural values

Plate 8.1 Urban flooding in New Zealand showing failures in hazard management
(photo by Neil Ericksen)

of riverine floodplains and shorelands could be preserved. In some cases, land-use approaches also included the evacuation of previously developed areas that were at risk from natural hazards. In New South Wales and New Zealand, especially, engineers in project-oriented agencies played a key role in arguing for greater use of land-use measures by local governments in their areas, as they recognized the fallibility of the structural works.

Although the non-structural provisions were often only augmentations of existing policy, putting them in place was fraught with difficulties. For example, in the United States the Carter administration ran into a political backlash when the president vetoed funding for pending flood control projects and attempted to increase attention to environmental values. As discussed in Chapter 4, efforts to institute restrictive land-use controls for flood-prone areas in New South Wales resulted in a political backlash leading to the present flood policy. New Zealand adopted a new policy aimed at greater technical and financial support for land-use planning for flood hazards in 1987, but that policy was over-shadowed by local government reforms. Elements of the policy were eventually incorporated into the *Resource Management Act*.

Nor is the evidence very strong that the past policy mandates have been effective in bringing about the desired changes. In the United States, the National Flood Insurance Program dating to the late 1960s was to become the main lever for inducing flood-prone communities to adopt floodplain building regulations in return for eligibility of federally subsidized insurance for property within flood-prone areas. The land-use planning component of this legislation was never invoked. Partly as a consequence of this failure, the insurance program became a different form of subsidy for development of floodprone areas; albeit with some adjustments to building practices.[3] More has been achieved through the planning policies of some states, as exemplified by Florida's strict development controls in coastal areas. In New South Wales, the highly prescriptive flood land-use policy of 1977 to 1984 had positive impacts on the environment by accident rather than by design. In New Zealand until the late 1980s, central government subsidies for structural controls, insurance, and relief, coupled with a strong pro-development anti-planning lobby, ensured only limited attention was given to integrated floodplain planning and management at the local level.

HAZARD MANAGEMENT CHOICES OF LOCAL GOVERNMENTS

These difficulties have led to the search for more politically sustainable approaches to addressing hazards. As documented in the first part of this book, Florida's growth management policy, New Zealand's resource management legislation, and New South Wales' flood policy each reflect different forms of policy and political learning about mandate design and policy implementation. The choices that local governments have made for managing natural hazards under the different intergovernmental mandates are examined in what follows.

149

Changing mix of strategies

In addressing choices by local governments about managing hazards, we consider the use of three measures: building regulations, land-use regulations, and structural protection measures. Building regulations address such items as flow heights in relation to flood levels, lateral strength, and anchoring of foundations. Land-use regulations address such things as allowable types of development, density of development, and protection of natural values. Structural protection measures include large-scale public works for control of flooding or coastal erosion, and smaller projects to address localized problems such as storm water run-off. These three types of measures are not mutually exclusive, since governments can employ a combination of building regulations, land-use provisions, and structural protection measures. Of interest is how that mix has changed over time, and in response to different types of intergovernmental policies.

The changing mix of local government choices about hazard management under the different intergovernmental mandates is shown in Table 8.1. The first two columns show the percentage of local government respondents reporting use of a given approach before and after introduction of either the New South Wales merits flood policy, the New Zealand environmental management legislation, or the Florida growth management policy. The last column shows the percentage of respondents who rate the respective policy as either very or somewhat important in bringing about the reported change in hazard-management emphasis. The relevant hazards differ for each setting. The New South Wales data address the efforts of local governments to deal with flooding with data restricted to those local councils that have participated in the merits floodplain management process. This restriction permits an assessment of the impact of the process on floodplain management choices. The New Zealand data address local councils' management of floods, coastal erosion, and earthquakes. Here again, the caveat applies that the local governments were only in the early stages of policy implementation at the time of our survey. The Florida data address the management of hurricane-related coastal flooding by local government.

The data show an increase in use of each type of measure with the exception of a decrease in use of structural protection measures in New Zealand. This suggests local governments are trying to achieve more balanced strategies for reducing flood hazards. The nature of the changes varies somewhat by type of measure as discussed in the following paragraphs.

Land-use regulations The data in Table 8.1 show that the largest changes in hazard management after the introduction of new planning policies are for land-use regulations. The changes are particularly marked for New South Wales and for Florida. It is interesting to note that the magnitude of change is similar under the cooperative policy of New South Wales and the coercive policy of Florida. Although some of the changes in New South Wales may be a

Table 8.1 Changing strategies for hazard management

Location and relevant hazards[b]	Percentage of local governments reporting use of measures[a]		
	Before most recent policy	After most recent policy	Percentage rate policy important[c]
Building regulations[d]			
New South Wales			
Flood hazards	71	98	68
New Zealand			
Flood hazards	90	92	25
Coastal hazards	84	82	27
Earthquake hazards	67	71	24
Florida			
Coastal flood hazards	100	100	–
Land use regulations[e]			
New South Wales			
Flood hazards	51	95	72
New Zealand			
Flood hazards	82	94	37
Coastal hazards	95	98	40
Earthquake hazards	49	54	29
Florida			
Coastal flood hazards	50	88	–
Structural protection measures[f]			
New South Wales			
Flood hazards	32	48	35
New Zealand			
Flood hazards	55	43	16
Coastal hazards	23	16	14
Florida			
Coastal flood hazards	75	95	–

Sources: Compiled by authors from surveys of local governments

Notes:

[a] Cell entries are the percentage of local government respondents reporting use of a given measure before and after the introduction of the New South Wales' flood merits policy, New Zealand's *Resource Management Act*, or Florida's growth management program.

[b] Relevant hazards are shown for each setting. The base number of cases is fifty-five for New South Wales, between forty-two and fifty-one for New Zealand depending on the hazard, and sixteen for Florida. Responses for New South Wales are restricted to those local governments that participated in the merits planning process.

[c] Percentage of respondents that rate the relevant policy as either "very" or "somewhat" important in bringing about the reported change. This item was not asked of Florida respondents.

[d] Controls on types of building or other requirements for building in hazard-prone areas.

[e] Restrictions or other rules governing development patterns in hazard-prone areas.

[f] Use of engineered works to reduce hazard potential.

151

hold-over from the prior policy that mandated strict land-use controls, the impact of the merits policy seems to have been an even more widespread adoption of land-use measures.[4]

While there is greater attention to land-use regulations in New Zealand under its new planning process, the movement from the old regime is less than for the other two settings. This reflects the fact that land-use controls were already clearly evident in local practice and in national legislation more than a decade prior to the *Resource Management Act*. In addition, land-use management formed an important element in the floodplain management policy of the National Water and Soil Conservation Authority just prior to its disestablishment in early 1988. These trends are borne out by the lower percentage of New Zealand respondents who rated the *Resource Management Act* as important for bringing about changes in use of land-use measures than the corresponding rating by the respondents of New South Wales.

The fact that coastal-related land-use controls in New Zealand rate so highly before and after the new policy was adopted is likely due to the National Water and Soil Conservation Authority adopting a tougher stand on structural controls on the coast much earlier than it did for flooding.[5] The low rating for earthquake hazards is expected since seismic zonation is still in its infancy in New Zealand. Because local governments are asked to address seismic potential under the *Resource Management Act* requirements, this situation is expected to change in the years ahead, as many councils are already having maps of seismic risks completed.

Building regulations Noteworthy changes in the use of building regulations for flood-hazard management are reported for New South Wales. Given the high percentages of local governments employing building controls prior to new policies, the shifts are minimal for New Zealand and not relevant for Florida. The universal use in Florida of building regulations for flooding reflects stringent federal requirements for building elevation as part of the National Flood Insurance Program. The high starting value for use of building controls for flooding by local governments in New Zealand reflects the fact that building elevation was a more widely used regulatory measure than were land-use controls under the earlier planning regime. The slightly lower values in New Zealand for use of building controls in addressing coastal and earthquake hazards probably reflect differences in awareness of building techniques for addressing these hazards. The *Resource Management Act* is not viewed by the respondents as an important stimulus for change in building regulatory practices since they view the *Building Act* as more important for these purposes (see Chapter 3).

Structural protection measures A key finding of Table 8.1 is that, with the exception of Florida, structural measures are relied on to a lesser extent in the mix of hazard management techniques than are non-structural measures. Moreover, in New Zealand the use of structural works is declining although relatively few

Plate 8.2 Destruction to older buildings in Florida from Hurricane Opal, 1995 (courtesy of Phil Flood, Florida Department of Environmental Protection)

respondents attribute this to the *Resource Management Act*. Rather, this change reflects other policies including: the diversion of funds from flood control works to irrigation projects during the 1970s; the recent economic policies of central government that slashed subsidies to all sectors; and the consequential shift in responsibility for funding new structural works and maintaining old ones since 1988 from central government to regional and local councils.[6]

The modest increases in reported use of structural measures for both New South Wales and Florida are due to the fact that national and state subsidies are still available for structural protection systems in these settings, and there is a strong history of such use. However, the types of works that are being constructed appear to be changing. For example, state officials in New South Wales report fewer large-scale projects and more small-scale projects being constructed under the merits planning regime. These small-scale projects fit more readily into the comprehensive and flexible approach to flood-hazard management of the merits-based policy than do large-scale projects. This is also consistent with the goals of the national Landcare Program in Australia, which seeks to have community and other groups provide stewardship over land. One other reason for this shift, which is debatable, is the observation that large projects that had benefit–cost ratios had already been undertaken by the mid-1980s thereby making it difficult to justify subsequent proposals.

Changing character of regulation

In addition to the mix of different hazard-management measures, it is important to consider the character of building and land-use regulations. These can range from highly specific rules, that leave little room for exception, to more flexible rules that permit variation in application depending on different circumstances. The general approach of Florida's growth management program is one for which greater certainty is sought in the application of local regulation, while also providing for accommodation of development needs – thereby promoting sustainable development. The explicit approaches taken by the merits policy of New South Wales and the *Resource Management Act* of New Zealand are flexibility in local development of regulations, and encouragement of local decisions that take into account local circumstances. As a consequence of these differences in policy mandates, we expect to find variation among settings in the character of building and land-use regulations.

The changing character of building and land-use regulations adopted by local governments is shown in Table 8.2. The key contrast is between use of strict and flexible measures. Strict measures are rules that prescribe particular building practices (i.e., specification standards) or land-use measures that restrict or otherwise prohibit development in hazard-prone areas. Flexible building regulations are rules that specify building performance standards, or criteria for making decisions about acceptable buildings. Flexible land-use measures are rules that permit low levels of development that meet certain criteria or allow other compensation (such as excavating to maintain flood storage capacity) for development in hazard-prone areas. As with Table 8.1, the comparison within each of these categories is between the percentage of local government respondents reporting use of a given approach before and after introduction of the new policy.

The results for the cooperative policies found in New South Wales and in New Zealand show a marked decrease in local governments reporting use of strict building and land-use regulations, and an even more marked increase in use of flexible measures. The changes are most marked for local governments in New South Wales that chose to participate in the flood merits planning process.[7] Strict rules for building and land use decreased from 48 to 25 percent and from 18 to 15 percent respectively, whereas flexible rules increased from 22 to 73 percent and from 33 to 80 percent, respectively. The change from strict to flexible controls is less frequent for local governments in New Zealand than for those in New South Wales. The lesser changes for local governments in New Zealand reflect the transition that they are going through from the old planning regime to the new regime.

The responses for local governments in Florida show a different pattern from that of New South Wales and New Zealand. This evidences a strengthening of floodplain management, even though strict building and land-use regulations still play a major role. Since introduction of the state's growth management program in 1985, there has also been a noteworthy increase in the use by local

Table 8.2 Changing character of regulations

Locations and relevant hazards[a]	Percentage of local governments with strict regulations		Percentage of local governments with flexible regulations	
	Before most recent policy	After most recent policy	Before most recent policy	After most recent policy
Building regulations[b]				
New South Wales				
Flood hazards	48	25	22	73
New Zealand				
Flood hazards	39	27	51	65
Coastal hazards	48	27	36	55
Earthquake hazards	36	21	31	50
Florida				
Coastal flood hazards	100	100	69	88
Land use regulations[c]				
New South Wales				
Flood hazards	18	15	33	80
New Zealand				
Flood hazards	22	22	61	72
Coastal hazards	27	9	68	89
Earthquake hazards	2	2	16	16
Florida				
Coastal flood hazards	44	50	50	88

Sources: Compiled by authors from surveys of local governments

Notes:

[a] Cell entries are the percentage of local government respondents reporting use of a given measure before and after the introduction of New South Wales' flood merits policy, New Zealand's *Resource Management Act*, or Florida's growth management program. Relevant hazards are shown for each setting. The base number of cases is fifty-five for New South Wales, between forty-two and fifty-one for New Zealand depending on the hazard, and sixteen for Florida. Responses for New South Wales are restricted to those local governments that participated in the merits planning process.

[b] Strict building regulations are defined as rules that prescribe particular practices (e.g., flood elevation requirements). Flexible building regulations are defined as rules that specify performance standards allowing for different means of adherence or rules that specify criteria for making decisions about acceptable buildings.

[c] Strict land-use regulations are defined as rules that prohibit or otherwise restrict development in hazard-prone areas. Flexible land-use regulations are defined as rules that permit low levels of development or other tradeoffs in return for development rights in hazard-prone areas.

governments of flexible measures. We attribute this to a general increase in understanding among planners, especially in the United States, of different tools for managing hazards. In Florida, flexible provisions are being added as an overlay to existing strict rules and regulations. Another interpretation, however, is that New South Wales and New Zealand are *changing* approaches, while Florida under its coercive mandate is strengthening its approach.

LOCAL APPROACHES TO HAZARD MANAGEMENT

The preceding data about the changing mix of hazard-management strategies and the changing nature of regulations evidence a greater sophistication in local government approaches to managing hazards than the practices of a decade or two ago. An important aspect of this is a shift in local government thinking about the appropriateness of building and land-use measures for addressing hazards. For New Zealand and New South Wales, the flexibility provided by their cooperative regimes is the means by which greater political acceptability of development management is being achieved. Local government officials in New South Wales, such as those from Fairfield in Sydney, who vigorously opposed land-use controls under the old regime now embrace more flexible land-use measures. The changes in hazard-management approaches also reflect learning by relevant professionals about the costs and effectiveness of measures for addressing the problem.

However, the learning and the changes in local government practices have not been the same for all local governments. One potential drawback of the cooperative regimes is that they offer a chance for local governments to do as little as, or even less than, they did under earlier planning regimes. This variation was evident in our discussion in the preceding chapter of gaps in procedural compliance under the cooperative regimes. The shift to flexible measures in New South Wales and New Zealand, versus the addition of new measures in Florida is also an indication of the potential for cooperative regimes to foster capture of local government policy by development interests. In the material that follows, we discuss a set of case studies of the flood-hazard management programs in Australia and New Zealand. The cases illustrate the variation in hazard-management practices under cooperative regimes, and reflect the different motivations that local governments have for substantive compliance (or non-compliance) with the goals of hazard management policies.

New Zealand cases

Although at the time of our study local governments had only a few years' experience under New Zealand's *Resource Management Act*, there is sufficient information about emerging directions to compare hazard-management approaches under the new and old planning regimes. Two cases were chosen: Invercargill, the southernmost city of New Zealand, which lies within the jurisdiction of the

Southland Regional Council; and Thames, a regional town in the Thames–Coromandel District, which lies within the Waikato Regional Council on the east coast of the North Island. Invercargill is indicative of a flood-prone community that seems to be using the flexibility provided by the resource management and building legislation to retrench from prior more comprehensive flood-management policies. In contrast, Thames is seeking a broader approach that varies the mix of structural and non-structural adjustments according to the nature of risk.

Invercargill case study   In the late 1970s and early 1980s, staff in the Southland Catchment Board expressed concern to city councillors about the build-up of flood hazard, to the point of raising issues about proposed developments with the Planning Tribunal. The potential for flood losses is evident from Figure 8.1.[8] It shows the built-up area in ten-year intervals to 1984 when a large flood covered the northern portion of the city. Over one-quarter of the built-up area had been created in the decade before the disaster at a time when planning regulations were being progressively tightened. Even with tighter regulations, compliance appears to have been limited. As a consequence, almost none of the

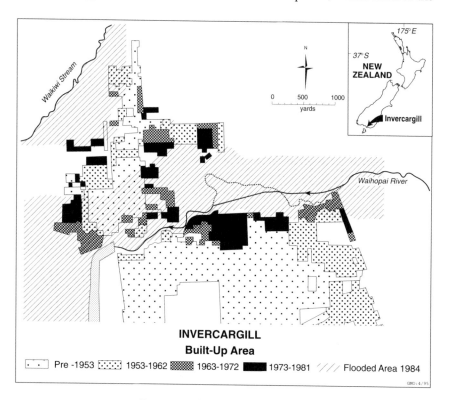

Figure 8.1 Flood-prone Invercargill

157

development in Invercargill was situated to avoid flood hazards. The losses from the 1984 flood exceeded NZ$50 million in a city of just over 50,000 people. The annual exceedance probability for the flood was estimated as being one-half of a percent, equaling the level designated a 200-year flood.[9]

At the time of the 1984 flooding, the National Water and Soil Conservation Authority was already reviewing its policy-bias towards subsidizing structural measures. In dealing with the Invercargill flood problem, the agency took a fresh approach that included planning options. In his report to Invercargill City Council, the city engineer cautioned councillors against using levees in densely populated areas. He recommended they adopt small upstream detention dams in the short-term, but limit the longer-term risk by planning for development away from the floodplain. The head of the regional catchment board supported this approach.

In March 1985, a comprehensive approach to the flood problem was adopted by the city council, making it a national leader in floodplain management. The council's district plan contained a commitment to pursue not only short-term flood control works, but also planning measures aimed at providing a long-term solution to the city's flood problems.[10] The planning measures included listing hazard-prone properties, re-zoning of hazardous land, minimum floor levels, and restrictions on subdivisions in hazard-prone areas. As required by law, these measures were publicly notified for objections and appeals, and some objections were heard by the Planning Tribunal. The National Water and Soil Conservation Authority enticed the city into employing planning provisions through an offer of a 70 percent subsidy for a combined detention dam and levee system. At the same time, the city had been deft in emphasizing the non-structural components in its district plan in order to cut the best deal.

Be this as it may, the manager of the catchment board warned at the time that: "as memory of the 1984 flood faded, city councillors and staff, along with property owners and developers, would exert pressure to further develop the floodplain – moves the board would resist." By 1988, this prophesy was realized as the hazard listing was not kept up to date and once again the city was considering proposals for development on the floodplain. Advocates of development argued that since so much money had been invested in protecting the land, it ought to be developed. Furthermore, the Ministry for the Environment did not follow through with the same policy when it took over flood policy functions in April 1988. In commenting about this, a planner in the Southland Regional Council ruefully remarked: " . . . although catchment authorities carried out their responsibilities for floodplain management – for example, flood plans, provisions in district schemes, publicity, education – the whole process ground to a halt because of a lack of implementation at national level; we feel let down."[11]

When Southland Regional Council took over from the Southland Catchment Board after the local government reforms of 1989, the structural works were almost completed. The new regional council therefore resolved to publicize the need for a planning approach to floodplain management. In 1990, the council staff wrote an article in the *Southland Times* that reviewed the history of decisions by all parties regarding floodplain management for Invercargill, and concluded:

> All local authorities must comply with specific statutory obligations aimed at controlling development on land which is or likely to be subject to flooding. These obligations or powers also carry a responsibility, and ultimately, a liability should a Council [e.g., Invercargill City Council] not exercise its duty in a responsible way. This liability is well established by case law in New Zealand.[12]

These statutory powers and responsibilities of local councils were incorporated into provisions of the *Resource Management Act* and the *Building Act* that came into effect in October 1991.

Interviews with individuals involved in these developments suggest three things. First the floodplain management policy of central government collapsed in 1988 at a critical point in the development of Invercargill's comprehensive floodplain management program. Second, pressure on council members from property owners to allow development of flood-prone land without restrictions had steadily increased as the protection works neared completion in 1988. Third, councillors who were committed to the original plan for comprehensive floodplain management had by and large left the political scene when the more flexible planning regime came into being in 1991. They were replaced by new councillors who were less knowledgeable of the original flood problem and its agreed-upon resolution and more open to the pro-development lobby. As a consequence of these several factors, the city council became more flexible in its interpretation of previous planning provisions. For example, dispensations for minimum floor-level elevations were given in some areas. To absolve itself from liability from future losses for such development, the council followed *Building Act* provisions for informing owners of flood-property about the risk and the liability they assume for future damages.

Reverting to a strategy of protection and warnings, rather than emphasizing the broader mix of measures that includes building and land-use regulations allowed for under both the previous and current planning regimes, gives effect to the catch-cry used by some of our interviewees to describe planning practice under the *Resource Management Act*: "educate don't regulate." For some, this means that councils should not prescribe what people must do to avoid flood hazards, but instead inform them of the problem so that they can make their own choices. This is a short-sighted approach to sustainable hazard management. It is well documented that, left to their own devices, most people will not take preventative action, such as elevating their land or buildings.[13] The threat of flooding may be too dimly perceived to warrant taking such action. The short-term benefits of developing a property without regard to the flood threat may outweigh the long-term costs. Or, property owners may not plan to own a given flood-prone structure for a long time, thereby reducing their risk of incurring a significant loss.

What is more, those who subscribe to the "educate don't regulate" approach fail to appreciate that to craft information and education programs – such as on flood hazards and ways of adjusting to them – requires a great deal of expertise. Properly done, such programs can effectively tap the attitudes of people who are predisposed to respond in appropriate ways. For others, the psychological barriers may be too great to effect any change – a theme taken up further in Chapter 10. Thus, if the "educate don't regulate" approach is taken to the extreme in Invercargill and elsewhere in New Zealand, it is hard to see how this type of flexibility will lead to more sustainable management of flood hazards on the nation's floodplains.

Thames case study A 1995 study of three district councils within the jurisdiction of Environment Waikato (renamed in 1993 from the Waikato Regional Council to emphasize its focus on environmental management) provides insights into different responses under the national policies instituted in the early 1990s.[14] One council adopted an approach similar to that of Invercargill in that it was emphasizing use of the *Building Act* when dealing with flood-prone development. The other councils, especially Thames–Coromandel District Council, were emphasizing the broader provisions of the *Resource Management Act*. Within the latter is the town of Thames, for which its flood risk is depicted in Figure 8.2. With a population of 6,000, the town is vulnerable to flooding from rivers, storm waters, and coastal storms. Serious floods occurred in 1979, 1981, and 1985 for which in total per capita losses exceeded NZ$2,000.

Located near the mouth of the Waihou River, Thames falls within the large-scale Waihou Valley Scheme that the central government agreed to fund in 1971. After the floods of the 1970s and 1980s, proposals for increasing protection from flooding were revised and slowly implemented. By 1995, almost all proposed flood control works had been completed. Although they appear to have reduced the frequency of flooding, a significant hazard remains, especially from the sea.

In order to build on the intolerance of flooding that has grown in the one-time sanguine community, and to avoid a false sense of security developing as protection works neared completion, the Waikato Regional Council decided in 1992 to prepare a plan to determine how best to manage the flood hazard under the then new *Resource Management Act*. Another reason for preparing the plan was that funds for future flood-control works and flood relief would not be available from central government, as they had in the past. Consistent with its proposed regional policy statement of 1993, the Waikato Regional Council (which by now had popularized its name to Waikato Environment) sought the cooperation of the Thames–Coromandel District Council in developing a joint flood-management plan for Thames.[15] The plan was spearheaded by a civil engineer and planner in the environmental planning section of Environment Waikato, and an experienced planning professional operating within a strong planning section in the Thames–Coromandel District Council. Their professional experiences were reflected in the principles that guided the plan they developed.

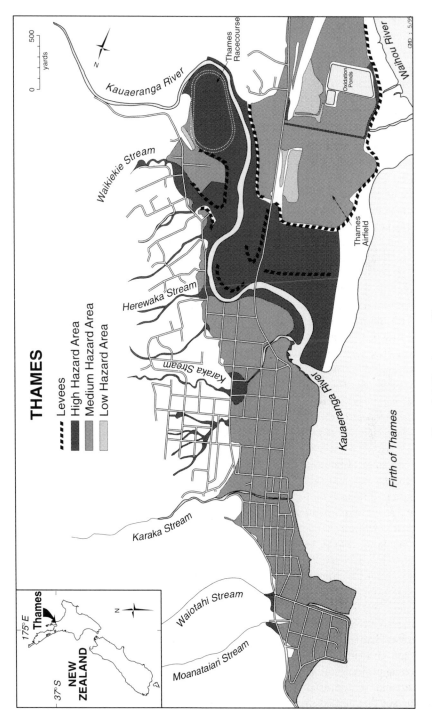

Figure 8.2 Flood-prone Thames

The main objective of the proposed plan was to "minimise the flood hazard without placing unnecessary restrictions on the rights of land-owners – in other words, to promote the sustainable management of the resources of Thames with respect to flood hazard."[16] The mix of measures that was devised constitutes a comprehensive but flexible approach to the flood hazard that reflects the spirit of the *Resource Management Act*. Three underlying principles guided the development of the plan's policies governing floodplain development. First, as the probability and/or severity of flooding increases, the level of flood avoidance (i.e., development control) should increase. Thus, additional building in the worst-affected areas is deemed inappropriate, while in other flood-prone areas, certain design criteria must be met before development would be allowed. Second, as the value of development increases over time, the level of flood avoidance should also increase so that potential flood damage is kept to levels that can be sustained by the community. It was also proposed that the district council would control planning for infrastructure in order to avoid having incremental changes (such as infilling and upgrading of existing facilities) undermine these provisions. Third, it was accepted that structural works which protect existing property are not in place to facilitate new development in the protected areas. The existing protection was to be maintained and extended where appropriate.

The outcome of this innovative approach to floodplain management will very much depend upon whether the process of consultation, currently under way, adequately empowers affected parties and engenders the necessary political will for adopting a comprehensive approach. There is potential for backsliding and placing heavy reliance on the enhanced structural control works, as appears to be what happened in the case of Invercargill. Convincing property owners and local politicians that a mix of measures is necessary is difficult to achieve given the appeal of structural works for protecting property that are paid for by someone else. A cause for hope is the learning that has occurred with respect to improving the process of consultation and participation in Thames as the planning staff and their respective councils have moved from the first to the second phase of the program. If inclusion of affected groups leads to their ownership of the proposals, it will result in a broad range of measures being tailored to suit the circumstances of particular zones within the flood-prone area. This innovation ought to result in a lessening of the flood hazard in the longer-term future, clearly a prospect that is encouraged by the *Resource Management Act*.

New South Wales cases

Unlike New Zealand's relatively recent changes in planning approaches, the merits-based cooperative flood policy in New South Wales has had nearly a decade to bring about change in local practice. Two New South Wales' local government case studies illustrate variation in responses to the merits policy and differences in approaches to addressing flood hazards. The Lismore case study illustrates the variable impact of the merits policy in bringing about change for a community that already had a strong set of non-structural flood-reduction measures in place. The Muswellbrook case study illustrates adherence to the merits planning process resulting in a set of comprehensive flood-management provisions.

Lismore case study Without doubt, Lismore is the most investigated flood-prone town in Australia. This dubious distinction results from its being located on low-lying land that has had sixty significant floods since 1860 – thirty-three of them in the forty years to 1980. Floods in 1955 and 1974 were in the order of 100-year events, but even smaller floods like that of 1989 resulted in flooding of business, industrial, and residential districts. The current flood hazard is depicted in Figure 8.3. Although much new development is now taking place in the flood-free area of Goonellabah, some 2,000 houses and 700 commercial buildings remain in the flood-prone area. The city's population is 40,000.

As documented in the following discussion, the merits-based flood policy of New South Wales has had an impact on how Lismore deals with floods, but it may not be positive in terms of flood-hazard management or even reduction of flood damages. The increased flexibility of the merits policy, which has been reflected in stages in Lismore, has arguably allowed more floodplain development, reduced the planning emphasis for flooding, and set the context for consideration of a new major levee scheme.

State officials of New South Wales first established an interdepartmental committee to examine the Lismore flood problem after the regional floods of 1945. Lismore introduced building controls for residential property in 1954. These required habitable floor levels for new developments to be 18 inches above the flood of record (close to the level of a 100-year flood). The state's inter-departmental committee eventually reported in 1958, recommending structural measures to protect existing property and the prevention of new development on flood-prone land. The structures were built by a newly created single-purpose Richmond River County Council (composed of representatives of the affected local councils) and were completed by 1974 at a cost of AUD$10 million. For Lismore, levees protected the central business district in the Browns Creek Basin and houses in South Lismore from the 10-year flood, which has a probability of 10 percent of occurring in any given year. Preventing further development on flood-prone land went largely unheeded, although the 1954 regulations for elevated habitable floor levels were generally enforced.

A large flood in January 1974 led to another state government interdepart-mental study that was completed in 1982.[17] The study recommended adoption of non-structural measures involving appropriate zoning and land-use regulation, but also including flood warning, property acquisition, and assistance with house raising. Levees were rejected on the grounds that they would be costly and have deleterious social and environmental effects. This emphasis on non-structural measures reflected the pre-merits coercive policy that was at the time the state's flood policy.

163

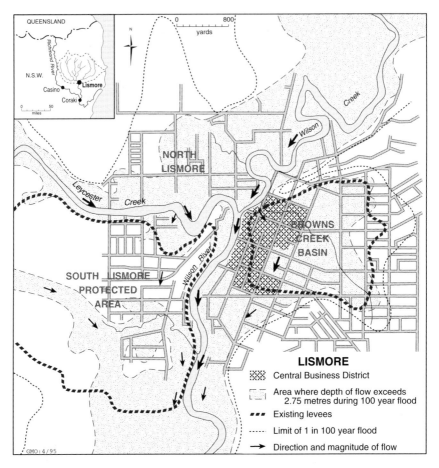

Figure 8.3 Flood-prone Lismore

In 1978, a program of voluntary purchase of homes was initiated by the Lismore City Council as an alternative to building a levee for North Lismore. This policy was later extended to the whole city. The state's report of 1982 identified 730 houses and 870 undeveloped blocks for acquisition at a cost of AUD$16 million (1982 dollars) finding that "this scheme is more attractive from an economic view and other aspects than major structural works."[18] Progress with property acquisition was slow, but sixteen houses and eleven blocks had been purchased by the mid-1980s. In contrast, house-raising by individual owners has a long history in Lismore. By 1979 over 90 percent of 2,000 dwellings at risk from the 100-year flood had raised their houses. This early and widespread response to flood risk is unique in Australia.

In the early 1990s, urban floodplain management in Lismore involved conflict between the proponents and critics of major engineering works as a solution to the ongoing flood stigma. This debate followed two major floods in 1987 and a modest flood and a false alarm in 1989. The 1989 April Fool's Day flood led the business community and some councillors to press again for major levees and a revision of the flood plan. The original consultants were re-employed to do a new flood study, published in August 1993, and an assessment of mitigation options, published in January 1994.[19] Whereas their original study stressed non-structural options and their benefits, the new study stresses structural measures and their costs. Of twenty-three possible options, two were selected by the city council in 1992 for consideration – one to protect all areas from the 20-year flood; the other giving additional protection to the central business district from the 100-year flood. After discussion and debate, the council selected the former, for which further evaluation was still under way at the time of our case study. Ironically, the debate about this proposal seemed to overlook the fact that the modest 1989 flood would have over-topped the proposed levees.

The preferred option can be seen as either addressing an equity issue or as a political compromise. The benefit–cost ratio of the proposed levee varies greatly from 2.15 for the central (business) area to 0.24 for North Lismore (residential). Protecting the latter will cost about AUD$50,000 per property, almost as much as it would for their purchase and clearance. The proposal appears to be designed to please both the commercial lobby in the central business district (led by prominent business people and various councillors) and resident action groups (such as the North Lismore Progress Association), who in contrast to South Lismore did not benefit from the flood protection system developed in the early 1970s.

However, even if the new scheme is accepted and government funding granted, the annual cost will amount to thirty to forty dollars (Australian) per taxpayer. The commercial sector's benefits are much greater, providing them highly subsidized flood protection. This fact was a concern to the opponents of major levees, which include the editor of the *Echo* newspaper, the Lismore Environment Center, and some councillors and business people, including an aspiring mayor. The levee scheme was destined to be a key issue in the local council elections in 1995. If the levee system is adopted, it would represent a major departure from the non-structural flood policies pursued over the last twenty years.

Three issues emerge under the New South Wales' merits flood policy as the Lismore debate unfolds. First, is the extent to which state agencies intervene in local decision-making. Partly in deference to the decisions of local elected officials, officials from the state's Department of Public Works have acknowledged the levee system as a viable option. A second issue is whether state or Commonwealth funding will be forthcoming, given the disparate distribution of costs and benefits between the residential and commercial sectors. A related issue is whether the Commonwealth government will provide assistance under the National Landcare Programme which favors innovative approaches to flood-hazard reduction rather than the use of structural measures.[20] The acceptable approach in Lismore to reduce existing damages is not clear, and one is left with the impression that the debate regarding structural and non-structural flood mitigation will be ongoing well into the next century.

The flexibility allowed by the state's merits-based policy in conjunction with several local factors seems to have contributed to the change in policy emphasis in Lismore. One factor is that the image-conscious and growth-oriented politicians and business people of Lismore want to protect their commercial interests. A second factor was a change in key personnel and resultant culture among a set of groups that include the professional and elected members of Lismore City Council, the regional office of the Department of Public Works in Lismore, and the Richmond River County Council. A third factor was clear state and federal political support for a levee – including an apparent endorsement by the Prime Minister. This has emerged because the area has become a swinging seat in state elections.

Muswellbrook case study   If the flexibility of the merits-based policy enables Lismore to change from a non-structural to structural approach, how is it that other communities can maintain and even strengthen a planning approach to flood-hazard management? This question is addressed by the Muswellbrook case study. The town was one of the first councils in New South Wales to complete the state's floodplain planning process under the merits policy. The resultant approach to managing flood hazards has been progressive.

Muswellbrook is a regional center, with a population of 10,000, in the Hunter Valley. The older parts of Muswellbrook have experienced floods on many occasions, most seriously in the region-wide flooding of 1955 and most recently in 1971 and 1978. Flood maps of the early 1980s show some 400 dwellings and nearly 100 commercial enterprises susceptible to a 100-year flood. The floodprone areas of Muswellbrook are shown in Figure 8.4.

Rapid economic growth in the Hunter Valley in the early 1980s affected Muswellbrook, which changed from an agricultural center to a sought-after residential location for employees of nearby coal and thermal power generation industries. Although planning approvals for new houses averaged ninety per year, there was limited pressure to build on the floodplain. Indeed, only a handful of buildings have been located on flood-prone land since formulation of the state-wide floodplain regulations under the pre-merits policy. There were various reasons for this. The floodplain was not as attractive for development as the higher parts of town which commanded a better view and where development infrastructure was provided by the local council. In addition, the council established a policy in 1983 that prohibited building in the 20-year floodway, and required new buildings between the floodway and 100-year flood area to be constructed of suitable materials and have habitable floor levels raised 18 inches above the 100-year flood height.

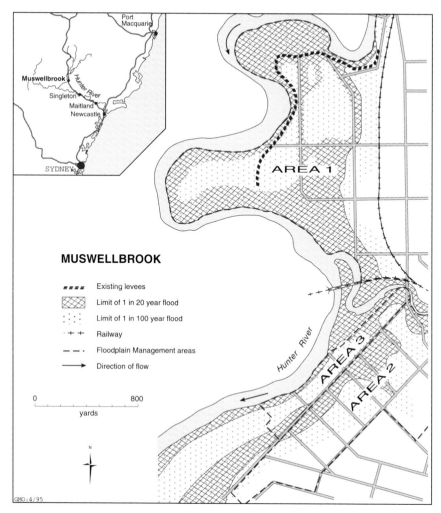

Figure 8.4 Flood-prone Muswellbrook

Although Muswellbrook had in place a comprehensive floodplain management policy restricting development within flood-prone areas in the early 1980s, it shifted direction in 1986 by embracing the more flexible development of these areas under the merits flood policy. The impetus for the change came from several sources. First, the professional local council staff were concerned with the unsatisfactory nature of the earlier planning regulations, and they also realized that extreme flood events posed a serious threat to residents. Second, staff within the regional office of the Water Resources Department, also located in Muswellbrook, wanted to see their detailed hydrological studies used for the development of local policies. Third, Muswellbrook officials were concerned over liability for approval of development in potentially flood-prone areas. In response to legal advice that having property owners indemnify the town was insufficient, a different approach was sought.

In 1984, Muswellbrook established a floodplain-management committee composed of both political and technical representatives. When the merit approach was promulgated by the state in 1985, the floodplain committee formally became the Muswellbrook Floodplain Management Committee. This group followed the merits planning process for which earlier hydrological studies were accepted as the equivalent of the required flood study, and the 100-year flood was adopted as the flood standard. Consultants were engaged to undertake a study of the social, economic, and ecological effects of flooding. This was later extended into a floodplain management study, as recommended under the planning guidance for the state's flood policy.[21] Discussions in 1985 between committee members and consultants established the overall aims for the study, which mirrored those of the state merits policy particularly in providing a more flexible approach to floodplain management.

The political nature of this planning approach was evident early on. A residents' action group formed as rumors of the solutions being discussed spread, and attitudes of locally based property lending agencies began to harden. To address these concerns, the floodplain committee called a public meeting to discuss the draft plan. This defused the situation and resulted in a member of the residents' action group being co-opted to serve on the floodplain committee. This action eventually increased the political acceptability of the plan, which was adopted by council in July 1988.[22]

Elements of the plan most relevant to Muswellbrook include voluntary purchase in Areas 1 and 3; house raising in Area 2; an L-shaped (untied) levee in Area 1; and an improved warning system (see Figure 8.4). These measures were estimated to cost AUD$2.8 million. The levees and warning system were proposed for development within a few years, and the other proposals were to be accomplished over a twenty year period. Other measures within the plan included flood education, emergency services, data gathering, linking river gauge heights to significant landmarks, and a process for future plan modification.

The levee was completed in 1992, thereby reducing over-floor inundation for two-thirds of the 104 properties in Area 1, and changing the risk for eighty-three of these dwellings from high to low hazard.[23] The most significant feature of the floodplain management plan is that the mitigation measures are vested in the town's broader planning documents, providing a firm legal basis for decisions about floodplain development. The relevant local environment plans have been amended to change the residential zoning of relevant areas, and these amendments have been approved by the state's Department of Planning. Details of the risk management measures for urban Muswellbrook are provided in three development control plans. Together with the local environment plans, these constitute key components for implementing the shire's floodplain management plan.

Although Muswellbrook had a modest flood problem and ample flood-free land for development, it became one of the first councils to adopt the merits approach and produced what state officials consider to be a model set of flood management measures. As discussed in this case, there are several interrelated reasons for Muswellbrook having become a leader in floodplain management. One set of reasons was a perceived need to address the problem of legal indemnification for flood losses that had emerged under its earlier flood policy. This was coupled with a clear vision among the town's professional staff of what was required to reduce the existing flood-hazard and to prevent problems from future development. A second set of reasons had to do with the politics of the planning process and resultant plan. The council established a floodplain committee that contained a range of stakeholders, thereby fostering acceptance of the plan. Its political viability was clearly enhanced by the fact that most people affected by the plan stood to benefit. A third set of reasons concerned the compatibility of the plan and the state's merits policy. By adhering to the state's planning process, Muswellbrook officials could make a strong case for state funds to help pay for the proposed measures.

CONCLUSIONS

This chapter has addressed the search for sustainable hazard-management strategies with a particular focus on flood hazards. In tracing the evolution of different strategies for addressing flood hazards in the United States, Australia, and New Zealand, we noted that bringing about sustainable approaches to hazard management has proven to be an elusive objective. Florida's growth management program, Australia's flood merits policy, and New Zealand's *Resource Management Act* reflect policy learning about intergovernmental regimes for inducing more balanced local efforts in hazard management. The promise of the latter two approaches is that these regimes provide the flexibility for local governments to craft policies that are better suited to local circumstances and politically more acceptable than those that could be obtained under prescriptive and coercive mandates. Given this promise, a central interest in this chapter has been depicting the changes that are occurring in local hazard management under cooperative regimes.

Our assessment of changes in hazard management by local governments in Florida, New South Wales, and New Zealand from the previous to the present planning regimes reveals similar overall trends in each setting. Structural controls for modifying hazardous events are less dominant than before. In each setting there is also greater attention to regulations concerning building practices and development patterns in hazard-prone areas as means for managing natural hazards. The key finding, however, is that the types of development and building regulations being employed by local governments differ between cooperative and coercive regimes. Local governments in both New South Wales and New Zealand have adopted flexible controls to a much greater extent than before, especially in

New South Wales where the cooperative merits-based policy has been in place for a decade. In contrast, local governments in Florida rely extensively on strict building and development controls accompanied by an overlay of some flexible measures. We attribute these distinctions in the character of local hazard-management measures to the flexibility allowed under the cooperative planning regimes found in New South Wales and in New Zealand. In particular, that flexibility permitted adoption of a mix of measures that was not possible under the prior restrictive planning regime in New South Wales and was less evident under the prior planning policy in New Zealand. The net effect of that flexibility, however, may be to open the gates once again to development of hazard-prone areas.

Our case studies of local hazard-management approaches in New Zealand and New South Wales suggest that while flexibility may be more politically acceptable locally than stricter controls, it has the potential to lead in different directions. Depending on the circumstances, local factors result in either a relaxed or rigorous application of policies for managing development in hazard-prone areas. The forces that railed against strict regulations in New Zealand and New South Wales are still evident under their flexible systems. Just as some communities dutifully applied strict regulations in the past, so too under co-operative regimes some communities appear to have developed comprehensive programs to reduce flood hazards. Yet, others appear to be doing little, in the age-old belief that growth – even in hazardous areas – is good for their community.

The cases suggest that the extent to which appropriate hazard management is undertaken under cooperative policies depends a great deal on the commitment of staff and elected representatives of local governments to hazard management, and on their political skills in persuading property owners that some forms of development regulations are in the wider community interest. Obtaining this requires political will and commitment that is not always present. We return to this challenge in the final part of this book.

NOTES

1 For an overview of these choices see Ian Burton, Robert W. Kates, and Gilbert F. White, *The Environment as Hazard*, 2nd edition (New York: The Guilford Press, 1993), pp. 31–65. A third strategy, not considered in this book, involves altering the distribution of losses from an event through use of insurance, tax codes, and disaster-relief provisions to redistribute the financial impact of disasters.

2 The seminal work in pointing out the human aspects of floodplain development is Gilbert F. White, *Human Adjustment to Floods*, Research Paper No. 29 (Chicago: University of Chicago, Department of Geography, 1945). The mixed experience within the United States in attempting to bring about a unified national program is reviewed in a 1992 report: Federal Interagency Floodplain Management Task Force, *Floodplain Managment in the United States: An Assessment Report*, 2 volumes (Washington, D.C.: Federal Emergency Management Agency, 1992). The Australian experience is summarized in David I. Smith and John W. Handmer, "Urban Flooding in Australia: Policy Development and Implementation," *Disasters* 8, no. 2

(1994): 105–117. The New Zealand experience is summarized in Neil J. Ericksen, *Creating Flood Disasters? New Zealand's Need for a New Approach to Urban Flood Hazard* (Wellington: National Water and Soil Conservation Authority, Water and Soil Miscellaneous Publication No. 77, 1986).

3 See Raymond J. Burby and Steven P. French with Beverly A. Cigler, Edward J. Kaiser, David H. Moreau, and Bruce Stiftel, *Floodplain Land Use Management: A National Assessment* (Boulder: Westview, 1985).

4 Among local governments in New South Wales that were not participating in the planning process, 22 percent of respondents report use of land use controls before the merits policy and 59 percent report use of them after the policy.

5 See: New Zealand, National Water and Soil Conservation Authority, "Policy on Natural Hazards and Limitations to Land Use," Circular 71/1/45, 5 February (Wellington: National Water and Soil Conservation Authority, 1982).

6 The trends in New Zealand for flood protection costs from 1950 to 1985 are presented in Neil J. Ericksen, *Creating Flood Disasters?*, op. cit., p. 204. Three reports that review flood-protection schemes acknowledge that many previous works were over-capitalized, and there were problems in maintaining existing systems without subsidies from central government. See: New Zealand, Ministry for the Environment, *Funding Options for Flood Protection: A Review of the Affordability of Flood Protection on the West Coast* (Wellington: Ministry for the Environment, 1992); Gary Williams, *Review of Flood Protection Scheme Maintenance – with particular reference to Canterbury and Hawkes Bay*, Report to the Ministry for the Environment (Otaki, New Zealand: G & E Williams Consultants Ltd, June 1994); and Barry Butcher, *Review of Flood Protection Scheme Maintenance – Bay of Plenty, Southland and Waikato Regional Councils*, Report prepared for the Ministry for the Environment (Wellington: Ministry for the Environment, 1994).

7 Among local governments in New South Wales that were not participating in the planning process, the use of strict building regulations increased from 48 percent of local governments to 58 percent, and the use of flexible rules increased from 3 percent to 26 percent of local governments. The corresponding increases in strict land-use controls were from 6 to 9 percent, and in flexible land-use controls from 16 to 50 percent of local governments.

8 Figure adapted from Neil J. Ericksen, *Creating Flood Disasters?*, op. cit., Figure 4.23, p. 65.

9 Dallas M. Bradley, *A Flood Risk Assessment for Invercargill City*, Publication No. 45 (Invercargill: Southland Regional Council, July 1993).

10 See: City of Invercargill, *District Scheme Plan Review* (Invercargill: City of Invercargill, 1989).

11 Interview, 20 February 1994.

12 Cited in: Southland Regional Council, "Floodplain Management," Memo to the Ministry for the Environment (Invercargill: Southland Regional Council, 28 March 1990).

13 For a review of relevant literature see: Thomas F. Saarinen, ed., *Perspectives on Increasing Hazard Awareness* (Boulder, CO: Institute of Behavioral Science, University of Colorado, 1982), and National Research Council, Committee on Risk Perception and Communication, *Improving Risk Communication* (Washington, D.C.: National Academy Press, 1989).

14 Sarah Chapman, *Dangerous Liaisons: Managing Natural Hazards Under a Cooperative Mandate*, unpublished Master of Social Science Thesis (Hamilton: University of Waikato, Department of Geography, 1995).

15 See: Environment Waikato, *Proposed Regional Policy Statement* (Hamilton: Environment Waikato, 1993); Environment Waikato, *Draft Thames Flood Management*

Plan Stage 1, Technical Publication No. 1993/2 (Hamilton: Environment Waikato, May 1993); and Environment Waikato, *Issues and Options Report: Thames Flood Management Plan Stage 2*, Technical Publication No. 1994/2 (Hamilton: Environment Waikato, May 1994).

16 David E. Ray and Bain A. Cross, "Flood Management Planning for Thames Under the RMA," *Proceedings of a Rivers and Floodplain Management Seminar*, 11–13 October 1994 (Lower Hutt: Works Consultancy Services Central Laboratories, October 1994), p. 2.

17 See: New South Wales, Department of Public Works, Richmond River Interdepartmental Committee, *Richmond River Valley Flood Problems*, Volume 2 (Sydney: New South Wales, Department of Public Works, 1982).

18 New South Wales, Department of Public Works, Richmond River Inter Departmental Committee, *Richmond River Valley Flood Problems*, Volume 2, op. cit., p. 2.

19 See: The Council of the City of Lismore, *Lismore Flood Study and Floodplain Management Study: Stage 1 – Flood Study* (Sydney: Sinclair Knight & Partners Pty Ltd, August 1993); and The Council of the City of Lismore, *Lismore Flood Study and Floodplain Management Study: Stage 2 – Testing the Flood Mitigation Options* (Sydney: Sinclair Knight & Partners Pty Ltd, January 1994).

20 See: Commonwealth of Australia, Department of Primary Industries and Energy, *Guide to Project Funding Under the Commonwealth State Component of the National Landcare Program 1994/95.* (Canberra: Department of Primary Industries and Energy, 1994).

21 See: Cameron McNamara Pty Ltd, *Social, Economic and Ecological Effects of Flooding at Muswellbrook* (Brisbane: Cameron McNamara Pty Ltd, September 1986).

22 All members of the original floodplain committee who were interviewed agreed that the public meeting and the inclusion of an action group member on the committee in 1985 was advantageous in having the final plan accepted. The floodplain management plan was adopted by council in July 1988, with little public dissent, as Shire of Muswellbrook, *Floodplain Management Plan, Shire of Muswellbrook* (Muswellbrook: Shire of Muswellbrook, n.d.).

23 In the Muswellbrook flood plan, a high hazard is defined as when the product of depth (metres) and velocity (metres per second) exceeds one. This in effect replaces the 20-year flood as the definition for a "floodway" under the state policy that preceded the merits policy.

9

EXAMINING OUTCOMES OF COOPERATIVE POLICIES

The preceding chapter demonstrated that the cooperative policies in New South Wales and in New Zealand have contributed to noteworthy changes in local approaches to managing development in hazardous areas and, by extension, to environmental management more generally. The hallmarks of those changes are increased flexibility for local decisions about development and more comprehensive approaches to addressing the risks posed by natural hazards. At a very general level, these changes reflect key tenets of those who argue that sustainability is fostered by appropriate local decisions about land use and development patterns.[1] Yet, the case studies of the preceding chapter provide a cautionary tale. For some of the cases, the local decisions appear to have resulted in increased exposure of residents to natural hazards, thereby calling into question the future sustainability of parts, if not all, of these communities.

This leads us to consider two challenges to cooperative intergovernmental policies governing local environmental management.[2] One challenge is that local government officials cannot be trusted to make socially beneficial decisions about land use and development in hazardous areas. Those who argue this perspective see these officials as either single-mindedly pursuing economic development, or dominated by parochial development interests. A second, related challenge is that there is an uneven playing field so that development decisions will not be fair because of capture of local regulation by a dominant set of interests. Those who argue this perspective see local governments either lacking needed expertise or having insufficient will to make appropriate decisions, and therefore being vulnerable to capture. The solution to either of these challenges is for higher-level governments to step in and exert strong controls over local government decisions, thereby reverting to coercive intergovernmental policies.

The critical test, of course, is whether cooperative policies are improving prospects for future sustainability. Assessing this is fraught with difficulties since what constitutes sustainable outcomes is debatable, particularly as it relates to risks posed by natural hazards. Although risk reduction is clearly desirable, it may not be an end in itself.[3] The costs of reducing risks may outweigh other benefits, there may be sufficient social disruption (e.g., in housing or employment) to make the costs of risk reduction prohibitive, or the risks may be sufficiently small to make them inappropriate to address. Nor is it clear that growth and development are desired ends, as they may unduly increase the exposure of people or property to disastrous natural events in future. The evolving consensus among researchers is that sustainability with respect to natural hazards consists of outcomes where the risks of catastrophic losses are reduced, while community resilience to less dramatic natural events is increased.[4]

This chapter considers the above challenges to cooperative policies and the outcomes of local development decisions under these policies. Two types of evidence are considered. One is the perceptions that local officials have of decision-making under cooperative policies and of changes in exposure of their community to hazardous natural events. Because reliance on such perceptions is fraught with problems, we emphasize the overall patterns in perceptions rather than attempting to make fine-toothed distinctions among them. Our second set of evidence consists of an evaluation of actual changes in exposure to flood risk for selected communities under the cooperative flood policy of New South Wales.

CHANGES IN DEVELOPMENT DECISION-MAKING

Those who say local governments cannot make socially beneficial decisions about development or who argue that such decisions will be unfair call into question decision-making about future development by local government officials. We consider the perceptions that officials have of the changes that cooperative policies have brought about in such decision-making. This includes changes in the flexibility that local governments have established for rules governing development and their willingness to negotiate with developers. A related consideration is officials' perceptions of the fairness of decision-making.

The perceptions that local government respondents have of changes in development decision-making since the implementation of cooperative policies, and the importance they attach to the policy in bringing about the changes are summarized in Table 9.1. For New South Wales, we show results for all councils, while recalling that a sizable minority of them chose not to follow the prescribed planning process. For New Zealand, we show responses for district councils, including unitary authorities. Because the *Resource Management Act* was in the initial stages of implementation at the time of our study, we expect to see less change for the New Zealand data than for the New South Wales data.

Flexibility and willingness to negotiate

The perceptions of decision-making are consistent with the changes in local development rules discussed in the preceding chapter. These perceptions are consistent with what we would expect of the merits basis of decision-making in New South Wales and with the shift in New Zealand from prescriptive rules to consideration of effects when developing new rules for development. The

Table 9.1 Perceived decision-making outcomes

Decision-making consideration and percentage reporting[a]	New South Wales local councils[b]	New Zealand district councils[c]
Flexibility in rules or regulations concerning development		
Increased flexibility	51	28
Decreased flexibility	7	10
Policy important for change	65	72
Council willingness to negotiate with developers		
Increased willingness	49	15
Decreased willingness	6	11
Policy important for change	67	65
Fairness of decision-making about development in hazardous areas		
Increased fairness	42	22
Decreased fairness	3	0
Policy important for change	70	49

Sources: Compiled by authors from surveys of local governments
Notes:
[a] Cell entries are the percentage of respondents from local councils indicating a given response. Respondents were asked whether there was an increase, no change, or decrease in an item under the mandated planning process. Respondents were also asked how important the relevant policy was in influencing the reported change. The "policy important for change row" is the percentage reporting "moderate" or "very" important responses.

[b] Flood-prone local councils, based on responses from 127 localities.

[c] District councils and unitary authorities, based on responses from fifty-nine localities.

New South Wales data demonstrate that local officials strongly perceive more flexibility in development rules and greater willingness to negotiate with developers or others who want to build in hazard-prone areas. For those councils that have substantially completed the merits planning process, the percentage of respondents reporting increased flexibility in rules, willingness to negotiate rules with developers, and fairness in decision-making is greater by some 25 percentage points for each item than for those who have not participated in the process. The New Zealand data also evidence increased flexibility, but as expected the changes are less prevalent than in New South Wales.

What constitutes flexibility in development rules and negotiation with developers? Our case studies in New South Wales suggest the locally derived regulations under the cooperative merits policy are more flexible than the state-wide regulations that they replaced under the prior coercive policy, in part because they relate better to local factors. The flexibility is further increased, as for example in Muswellbrook, when multiple development control plans are produced that detail regulations for more localized areas. Once the new regulations are in force, they are the rules to be followed. Negotiation typically takes place over the means for adhering to the rules, revolving around best practices to meet the locally specified standards, rather than over the standard per se.

Understanding variation in the willingness of local government to negotiate adherence to development rules provides clues about the potential capture of development decisions. To get at this we undertook a statistical modeling for the New South Wales data of survey respondents' perceptions of changes in willingness to negotiate with developers or other groups. For this analysis we drew a contrast between those councils that reported increased willingness to negotiate and those that reported no change or decrease in willingness, and then statistically identified those factors that account for this contrast. We present the results of the logistic regression in an appendix. Of more interest here are the substantive findings.

Our modeling shows that two factors play key roles in willingness to negotiate the terms of development. The most important factor is the flexibility of development rules, which we interpret as being critical in structuring opportunities for negotiation. A second important factor is the degree of staff professionalism (measured as the percentage of staff with professional certifications). Councils with greater degrees of professionalism are more likely to be willing to negotiate with developers. This makes sense since professionalism is a key to making

Plate 9.1 Contrast between old and new requirements for elevation of land and buildings in South Huntley, New Zealand (photo by Neil Ericksen)

judgments about the merits of development proposals. Professional judgments are also easier to defend from political attacks in arguing that such judgments are based on a standard of professional practice.

Two points are noteworthy about other factors that did not show up as influences in our statistical modeling. First, political factors do not appear to play a role in influencing local governments' willingness to negotiate. In particular, we do not find evidence that the development community influenced the decision-making of local government officials in their willingness to negotiate compliance with their rules. This contradicts the notion that cooperative policies favor development interests. The second key point is that as the amount of hazard-free land increases, there is less tendency to want to negotiate development outcomes. This presumably reflects the fact that under these circumstances there is less need to negotiate about development in hazard-prone areas. The preferred alternative is to divert development to hazard-free locations.

In short, these findings suggest that willingness to negotiate is structured by the rules of the game, the skills of the players, and available alternative options. None of this shows capture by particular interests, but it does leave the door open. In particular, we surmise that pressures to allow for development in hazard-prone areas will be greater in localities with flexible development rules, limited alternative sites for development, and less capable staff. We return to these later in this chapter.

Fairness of development decisions

The perceived fairness of development decisions is relevant because it speaks to one of the challenges to cooperative policies, as well as to the longer-term viability of such policies. The increased flexibility of decisions and willingness to negotiate with developers, or others, could be interpreted with alarm by those who subscribe to the view that local government officials are often captive of local development interests. Without knowing the specific outcomes of this new flexibility, it is hard to say whether these data support the challenge that cooperative policies "give away the store" to development or interests looking to exploit valuable land in hazardous areas. If one interprets fairness as balanced decision-making, the evidence in the last part of Table 9.1 suggests that this is not the case. The respondents from New South Wales clearly perceived an increase in fairness of decision-making and attribute it to the merits policy. To a lesser extent, the respondents in New Zealand also perceive an increase in fairness.

To understand variation in perceived fairness, we developed a statistical model for the New South Wales data that drew a contrast between those respondents that reported increased fairness and those that reported no change or a decrease. (The statistical results are also reported in the appendix.) The process is perceived as being fairer where local governments place greater emphasis on managing development (i.e., take the process more seriously), have increased flexibility in rules, and are more willing to negotiate development outcomes. Increases in staff are also positive influences on the perceptions of increased fairness.

In essence, the respondents in New South Wales believe that key principles of the merits planning process – increased flexibility and willingness to negotiate on the merits of development proposals – are central to a fairer system. Where developers have stronger influence, respondents from those governments perceived less fairness in decision-making. The latter finding is, however, arguable in that the results fail to meet conventional statistical tests. Nonetheless, the finding suggests that capture and an associated perception of unfairness occur in some circumstances.

CHANGES IN PERCEIVED EXPOSURE

The key test of environmental management for hazardous areas is the extent to which future sustainability is being fostered. We characterize this in what follows as a discussion of the change in exposure of people and structures to hazardous natural events. Of particular interest is the increased exposure to large events that exceed those that are normally planned for (e.g., the 100-year flood). We consider sustainability to be enhanced if the risks of such catastrophic losses are reduced, while community resilience to less dramatic natural events is increased. Our interest here is how such exposure has changed under the cooperative policies in New South Wales and in New Zealand.

Like findings in the previous chapter that localities employ a mix of hazard-management solutions, the evidence so far in this chapter suggests a mix of outcomes with respect to exposure. As shown in Table 9.2, the majority of respondents in New South Wales and in New Zealand thought there was no change in exposure brought about by actions undertaken with the cooperative policies. Sizable percentages of respondents were, however, split in their perception of the direction of change.[5] The fact that not larger percentages of officials saw the policies as important forces in bringing about risk reduction suggests the policies in New South Wales and New Zealand are not viewed exclusively as mandates to reduce risk. Some respondents felt that the price of balancing different objectives is increased exposure to risk. The New Zealand data also undoubtedly reflect the newness of the policy.

In comparison to these findings, respondents to our Florida survey saw their more coercive and prescriptive policy as a strong mandate for risk reduction. Eighty-seven percent of the respondents to our survey of local governments in Florida reported that state requirements to address natural hazards were important in the degree of attention their government gave to reducing such risks. Seventy percent thought that one result of following the mandated planning process was increased attention to reducing the amount of development in hazardous areas.[6]

To understand variation in perceived exposure under cooperative policies, we developed a statistical model for the New South Wales data that drew a contrast

Table 9.2 Perceived exposure outcomes

Percentage reporting[a]	New South Wales local councils[b]	New Zealand district councils[c]
Increased exposure	16	10
Decreased exposure	22	8
Policy important for change	56	42

Sources: Compiled by authors from surveys of local governments

Notes:

[a] Respondents were asked whether there was an increase, no change, or decrease in exposure of people and property to hazards under the mandated planning process. Respondents were also asked how important the relevant policy was in influencing the reported change. The "policy important for change" row is the percentage reporting "moderate" or "very" important responses.

[b] Flood-prone local councils, based on responses from 127 localities.

[c] District councils and unitary authorities, based on responses from fifty-nine localities.

between those councils that reported increased exposure and those that reported decreased exposure. (The statistical results are also reported in the appendix.) Of particular interest are the effects of the different influences on development decision-making discussed in the preceding section. The modeling suggests that respondents believe increased flexibility in rules governing development contributes to increased exposure. In addition, the likelihood of increased exposure is thought to be greater when developers have strong influence in decision-making by local governments. Yet, exposure is decreased as local officials are more willing to negotiate with developers and as the capacity of local governments to manage flood hazards increases. Stated differently, the exposure-increasing influences of flexibility and developer influence are offset somewhat by the exposure-decreasing influences of negotiation and capacity. Not surprisingly, the likelihood of increased exposure is higher among those places with higher demand for development in hazardous areas and places with greater population growth.

These findings provide partial support for the concern that cooperative policies open avenues for the development community to capture decision-making about development, thereby leading to increased exposure. The findings suggest that there is less likelihood of increased exposure among those places that are serious about hazard management – those places that are willing and able to negotiate with developers. Respondents appear to believe that they can use negotiation as an effective means for obtaining outcomes that do not increase exposure. This is facilitated when there is hazard-free land available to accommodate new development. Being serious about hazard management presumably means that local governments work smarter at managing development in hazardous areas thereby decreasing, or at least preventing increases in, exposure.

When the efforts of local governments are weak, however, and negotiations

179

break down, the avenues for the development community to capture decision-making seem to open. The potential for capture is greatest among those governments with increased flexibility in their rules, for which developer influence is strong, for which development demand is strong, and for which there is no available hazard-free area to divert development. In short, the exposure outcome appears to depend largely on the willingness and ability of local officials to appropriately apply flexible rules and negotiate outcomes. Commitment to sustainable outcomes, and to a lesser extent capacity for managing development, loom large in this equation.

CHANGES IN HAZARD EXPOSURE AND SUSTAINABILITY

The preceding section evidences perception in a minority, but still sizable number, of localities of an increase in exposure of people or property to risks from hazardous events after introduction of the cooperative policies in New South Wales and in New Zealand. We attribute this to differences in decision-making contexts and outcomes. However, as noted earlier in this chapter, increased risk in itself need not signal unsustainability. The primary issues are the extent of exposure to catastrophic events, which by definition are unsustainable, and the extent of resilience of structures (and communities) to less catastrophic events. This requires a finer assessment of exposure than the broad-based perceptions elicited by our surveys.

In principle, it is relatively straight-forward to assess changes in exposure. One can look at development patterns within hazard-prone areas to see what has occurred and where development has shifted, if at all, under a given policy. One can also assess the extent to which development that existed in hazard-prone areas was either removed or strengthened in order to resist damage from natural events (e.g., elevating structures above flood stages, seismic strengthening). A more sophisticated version of this assessment would also consider the counterfactual of what might have occurred had the policy not been in place.[7] In practice, such comparisons are complicated by difficulty in judging the counterfactual and in linking development patterns to particular policies. In addition, it is difficult to gauge exposure to catastrophic events.

We address changes in exposure through selected case studies in New South Wales. The cases were selected in order to provide a contrast between "leading" and "lagging" jurisdictions with respect to management of flood hazards and compliance with the merits policy. (Details of case selection are provided in the methodological appendix.) An additional consideration was the availability of data about exposure of structures to floods prior to the merits policy. Given the selection procedure, the cases demonstrate a range of experience under the cooperative flood policy in New South Wales, including one case that illustrates non-compliance and others that illustrate different levels of effort to address flood hazards.

Plate 9.2 Devastation from Cyclone Tracy in Australia, 1974 (courtesy of the Disaster Awareness Program, Emergency Management Australia)

For each of the case studies, the experience with local floodplain management was reviewed and local stakeholders were interviewed. We assembled data about development patterns involving an inventory of flood-prone residential structures (below the level of the 100-year flood event). In most cases, we were able to compare this inventory with parallel inventories undertaken prior to the implementation of the merits flood policy.[8] Although the dates of the prior inventories vary from place to place, they provided a unique opportunity to investigate the changes in exposure patterns after introduction of the cooperative flood policy.

Contrasting outcomes: leaders and laggards

Table 9.3 shows, for the cases in New South Wales, our assessments of changes in exposure and characterizations of flood-hazard management efforts. By virtue of having profiles of flood exposure that change from one of increasing exposure prior to the merits policy to one of decreased exposure after the policy, Fairfield and Muswellbrook are considered leaders in flood-hazard management. In contrast, Liverpool is considered a laggard as evidenced by increasing exposure, and a history of ignoring the problem. Lismore occupies an intermediate position in that there is extensive flood exposure with little apparent change over time. As discussed in Chapter 8, Lismore city officials have been trying to find solutions to flooding for many years. Singleton represents a different intermediate situation

Table 9.3 Changes in exposure for local governments in New South Wales

	Selected New South Wales' councils				
	Fairfield	Muswellbrook	Singleton	Lismore	Liverpool
Flood plan(s) adopted under merits process	Yes	Yes	Yes	In process	No
Demographics					
Population (1991)	185,000	15,000	19,000	40,000	105,000
Annual growth rate (1981–91) percentage	2.7	.8	2.2	2.4	1.1
Pre-merits exposure					
Buildings at risk[a]	1,280 R 170 C	420 R 60 C	2,300 R 340 C	1,900 R 680 C	730 R 340 C
Change in exposure over prior period[b] (percentage)	Increase 4	Increase 8	Data not available	Data not available	Large increase 34
Post-merits exposure					
Risk management approach post-merits[c]	Non-structural provisions and minor structural works	Comprehensive	Levee and limited non-structural provisions	Non-structural and limited levee system	Limited effort
Change in exposure prior period to 1994[d] (percentage)	Decrease: –8	Decrease: –10	Increase: 13	Little change	Large increase: 33

Sources: Compiled by authors from case studies of development patterns and inventories of buildings

Notes:

a R = residential buildings; C = commercial buildings. Calculated from inventories of buildings

b This shows the percentage increase or decrease of buildings within the 100-year flood area during a period prior to the merits flood policy. Numbers are rounded.

c Non-structural approach defined as a use of a mix of property acquisition, warning systems and awareness-building actions. Levee defined as construction of a levee structural protection system. Comprehensive approach defined as use of a levee or other major structural work, house raising, and non-structural actions.

d The percentage increase or decrease of residential structures within the 100-year flood area subsequent to the merits policy of 1985. Calculations based on re-surveys of buildings undertaken by the authors in 1994.

in that officials completed the flood planning process, but the outcome – due to the topography of the area – was still one of increased exposure from catastrophic flooding.

The cases with successful outcomes evidence strong commitment to the merits policy, marked by the multi-faceted approaches to flood-hazard management discussed in the preceding chapter. For Muswellbrook the drive for action was from professional staff of the council who regarded the new merits planning procedure as an opportunity to clear up the issues of indemnity from flood losses and as an opportunity to obtain funding to provide protection for existing flood-prone development. The flood-management strategy included construction of a levee system accompanied by programs for house raising, property acquisition, and public awareness of flood hazards. For Fairfield, as elaborated on later, the impetus was the occurrence of two severe floods in quick succession in 1986 and 1988. The emphasis is on voluntary acquisition and paying the full cost of house raising.

The cases with less successful outcomes entail different reasons for reluctant action in implementing the merits policy. As elaborated upon later, new development and infill of existing flood-prone areas has continued in Liverpool, without noteworthy change, under both the coercive pre-merits flood policy and the cooperative merits policy. The flood risk is well defined, but this has not placed flooding on the agenda of the council, professional staff, or the community. The situation in Lismore, discussed in the previous chapter, is one of high awareness of flood risks with action stultified over the technical issue of whether or not structural protection through an expanded levee system is feasible. Currently a scheme to give limited protection to commercial and residential sectors is under debate. This is the only case study in which the commercial sector has played a major role in determining the form of the flood plan.

Singleton City officials embraced the flood merits policy as an opportunity to advance progress in implementing prior flood-hazard management plans, which were accepted as fulfilling the preliminary stages of the prescribed merits planning process. Given the topography of the area, levee protection is the only viable option for addressing flood risks short of moving the whole town. Indeed, Singleton is probably the largest community in the state that is protected by levees. Construction of an upgraded levee system in the late 1980s, combined with floor-level height minimum requirements are seen by the community as successful ways of building resilience to flood losses.[9] Because growth continues in the protected areas (as evidenced by our structures survey), changes to the level of exposure are debatable.

As illustrated by these cases, the relationship between exposure to floods and sustainability of communities is not straight-forward. Liverpool has undoubtedly increased exposure and thereby reduced future community sustainability. In contrast, Fairfield has reduced exposure and thereby increased long-run community sustainability. It can be argued that permitting residential development in any flood-prone location, within the limits of the probable maximum flood, decreases

183

long-term sustainability.[10] Development protected by structural measures such as levees is the prime cause of this concern, and as a consequence the long-term sustainability of such communities is now recognized as a potential problem. The extent of this problem depends on the design of the levee system and hydrological factors.[11]

A tale of two cities

These contrasts in outcomes are more fully illustrated by comparing the responses of Fairfield and Liverpool to the merits flood policy. The populations of Fairfield and Liverpool place them among the larger local government authorities in New South Wales. Both have considerable areas of development located on the floodplain of the Georges River and its tributaries in inner western Sydney. The headwaters of the Cabramatta Creek are situated in the jurisdiction of Liverpool city council with the lower reaches in Fairfield city council.

As noted in the preceding discussion, the approach of these two councils to floodplain management presents a juxtaposed contrast unmatched elsewhere in the state. Fairfield has completed its floodplain management plan and has an active program to reduce the risk to existing flood-prone development which is among the best in the state. Liverpool has yet to commence floodplain management under the merits system, and as such it represents a laggard in the process. The two cities also demonstrate the limitations of discussing flooding for the total area administered by a single authority since sub-areas present different problems and require different solutions.

Liverpool case study For at least the last twenty years, the approach of Liverpool to floodplain planning has been one of near neglect. This spans both the coercive and cooperative phases of state flood policy. Despite the council's large size, very few resources – either of professional expertise and time or direct funding – have been devoted to the flood problem. A floodplain management plan was produced in 1987, but it follows none of the preliminary stages prescribed in the state flood policy. The plan's short length and limited content do not compare with the detailed studies and plans produced by other local governments that were subjects of our case studies.

Despite these limitations, there are marked differences in exposure outcomes among the flood-prone developed sub-regions of the city, several of which are liable to flooding from the Georges River. The most positive outcome is the Moorebank area, an acknowledged area of risk that has experienced damaging floods in recent years. After studying the flood hazard in this area in the early 1980s, the state Department of Public Works recommended a voluntary purchase program for flood-prone structures and participated in funding the scheme. Our data indicate that in 1986 some 150 residential properties were at risk for this area, and by 1994 nearly forty had been removed and seven others had been raised in height. The area vacated had been converted to open space for recreation. Acquisition of flood-prone properties later slowed because Liverpool's councillors were unwilling to commit the necessary funding to take advantage of available state and federal assistance.

Three other developed sub-regions within the Liverpool area show increased development in flood-prone areas. The most striking of these is the continuing development of high-value residential dwellings in the Chipping Norton area. This stretch of the Georges River floodplain was reclaimed from gravel extraction ponds in the mid-1980s to form an area for new development. We estimate an increase of over one-third of the number of dwellings at risk since 1986. The council encouraged development in this area by adopting a lower standard for flood risks, apparently because development pressures were particularly strong for this area. The other two areas of development, the Warwick Farm and Heathcote Road areas, also located on the Georges River floodplain, are much older developments. New dwellings are dominantly of an infill nature or redevelopment of individual lots for which we estimate a 15 percent increase in the number of residences in the flood-prone areas since 1986.

Another flood-prone area is associated with a large new residential development in the Upper Cabramatta Creek catchment, a minor tributary of the Georges River. Planning for this new land release has included a thorough study of the changes in flood discharges and frequency that will occur as a result of the construction of houses and infrastructure. The natural watercourse will be preserved and flood detention basins and on-site detention measures for individual dwellings are designed to restrict the flood discharges to those of the pre-development situation. This reflects recent state policy for development in small catchments and the aims of total catchment management.

The outcome of the Liverpool council's inaction is that the construction of new dwellings in flood-prone areas along the Georges River continues. This applies to major developments in the Chipping Norton area as well as to infill at Warwick Farm and Heathcote Road. It is also clear that awareness of the community at risk is extremely low. There has been no effort by the city to inform or educate the community about flood risks, no development of emergency management procedures, and no development of a locally based flood warning system. In addition to the increased number of residences, there are exceptionally large areas of commercial and industrial activity in the flood-prone areas.

The surprise is that the Liverpool City Council has succeeded for so long in ignoring its responsibilities, and that state agencies have not been more aggressive in attempting to remedy the situation. The case represents the problems posed by inaction. Despite the lack of a comprehensive floodplain-management plan that is called for under the state flood policy, financial assistance has been forthcoming for the Moorebank area because plans for addressing this problem were under way prior to the new policy. Staff of the state's Department of Public Works were active in providing technical assistance and pushing to have this flood problem addressed.

The Liverpool case also illustrates the consequences of a lack of council or staff commitment to addressing flood hazards under either cooperative or coercive state mandates. One reason for the limited commitment is that, Moorebank apart, none of the flood-prone areas have experienced a damaging flood since 1956. The years 1993 and 1994 saw major changes in the organization of the Liverpool City Council, including the appointment of new senior staff in engineering and planning. The previous lack of attention to floodplain management has subsequently been acknowledged by the professional staff. In mid-1994, the city council formally approved the establishment of a floodplain management committee, which constitutes the first step in the state's recommended planning process under the merits policy.

Fairfield case study This case provides two key contrasts. The experience is a direct contrast with Liverpool in that Fairfield's elected representatives embraced the state's new flood policy and used planning measures to reduce exposure to floods and enhance community sustainability. In addition, this case provides a contrast between experiences under the previous coercive state policy and the experiences under the more cooperative merits policy. Flooding for much of Fairfield is from Prospect Creek and its upstream tributaries. The southern margin of the area is exposed to flooding from the main Georges River. The timing of the major floods is similar to that for Liverpool, but flood damages in the 1980s have been more extensive. Fairfield suffered extreme floods in 1986 and again in 1988. Prior to the 1980s, the last significant flood was in 1956 which, like Liverpool, predated the major residential development.

The saga under the prior flood policy is one of compliance, followed by resistance as the reality of that policy set in. The Fairfield City Council adopted in February 1981 a policy limiting development and use of flood-prone lands. This policy was generally enforced and only a handful of dwellings were constructed below the 100-year flood line between 1975 and 1986. However, the flood problem and flood awareness were low-priority issues for both the council and the community. In November 1982, the state Water Resources Commission (later renamed the Water Resources Department) released draft floodplain maps for most of the Fairfield area. These were the first flood maps for a developed area in metropolitan Sydney. The council considered that it was their duty to inform, by letter, all of the residents located within the limits of the 100-year flood. This action led to an extensive, adverse community reaction. The concerns were that the maps would lead to depressed house values, limit the ability to obtain financing for house purchases, and have adverse effects on the availability of insurance (which was erroneous as flooding has always been excluded from such policies in New South Wales).

To press these concerns, the South Western Flood Action Group was formed. Interviews with a number of the key committee members of the action group confirmed that their opposition was based on the disbelief that Fairfield was flood prone, and as such they thought the maps were "plain wrong." The action group organized public meetings (some with attendance close to 1,000), played a key role in mobilizing written submissions in response to the public release of the draft flood map, and actively lobbied leading government politicians for changes in the policy. The intense local feeling continued unabated until a state election was called in early 1984. As discussed in Chapter 4, the state election was the catalyst for policy change. Labour Premier Wran announced, days before the election, that the flood maps would be withdrawn throughout the state and that a new policy would be formulated.

Dispute over the technical aspects of the flood map engendered a series of additional hydrological studies, and a series of state-sponsored studies and proposals for addressing flood risks. Debate over these proposals was overtaken by the severe flood of August 1986. This approximated a 20-year flood-event along the Georges River and was a more extreme, 100-year event for Prospect Creek. Close to a thousand homes in the Fairfield area were flooded, and over 500 of these experienced over-floor inundation. The action group called a public meeting attended by some 600 angry Fairfield residents. The Deputy Premier and Minister of Public Works, Jack Ferguson, who was a strong advocate for floodplain management, attended the meeting. Following the meeting, the Fairfield council established a liaison committee which included representatives from the action group. This was the direct forerunner of the floodplain management committee, which is the recommended first step in the state's floodplain planning process.

The 1986 flood also provided valuable calibration for hydrological studies, and as a follow-up further studies were undertaken by consultants. These included a detailed assessment of the flood damages from the 1986 floods, floodplain-management studies for three main sub-catchments, and the completion of the Georges River hydrological study. Further opportunities for calibrating hydrological data and renewed community pressure arose after the floods of April 1988. These were of equal magnitude to those of 1986.

Toni Lord, a flood victim, contested the council elections in 1987 and was elected, based on a campaign that was based on the single issue of local flooding. Councillor Lord has remained on the council, serving for a term as mayor in the early 1990s. This has ensured that flood issues remain firmly on the council agenda and the other fourteen councillors, many with the same political affiliations, respect her expertise and views on such matters. Councillor Lord has served on the city's floodplain management committee since its inception, including a term as chairperson of the committee.

The outcome of the severe floods of 1986 and 1988 and resultant political action was the forging of closer links between the council, community, and the earlier berated staff of the state Water Resources Department. The council appointed a flood engineer who, by the mid-1990s, was granted substantial staff resources devoted solely to flood issues. The floodplain management committee met monthly and the various studies, all of a high quality, formed the basis for the floodplain management plan that was recommended under the state's flood policy. The final stages of this were formally accepted by council in 1992.

The plan called for a mix of minor structural works and non-structural approaches to flood-hazard management. For the upper catchments, the mitigation options selected were a combination of detention basins, channel modification, and minor local levees. For the remaining sub-region, Lower Prospect Creek, the mitigation options were for voluntary acquisition and house raising. This policy has been vigorously pursued, and by mid-1994 our surveys indicate that fifty-five houses had been acquired and a further seventy-three raised. These actions reduced residential exposure to flooding by some 15 percent for this area of Fairfield. This is very fast progress for measures that require acceptance by individual home owners. These actions contribute to a set of outcomes that both increase community resilience to floods and contribute to longer-term sustainability.

Three key factors contributed to the acceptance and implementation of Fairfield's flood-management programs under the state's merits-based flood policy. First, is the perceived equity of the programs involving all flood-prone residents. To demonstrate this, the council produced a map in leaflet form, distributed to all householders, that showed the proposed action for individual houses (no action, acquire or raise with the height to be raised given). Second, is the unprecedented availability of financial assistance, shared by all levels of government, that followed the development of the flood-management plan. Third, is the commitment to flood-hazard management of Fairfield's elected officials and active citizens. A combination of an active, high-profile council and the lobbying efforts of the Fairfield's citizen-based flood organization has led to a series of visits from federal government ministers with responsibility for flood issues and, more recently, new funding. Clearly, the extreme floods of 1986 and 1988 contributed to increased awareness of a problem.

CONCLUSIONS

The examination in this chapter of the perceived outcomes of cooperative policies and case studies of changes in profiles of exposure to floods in New South Wales serve as a limited test of two challenges that have been posed about cooperative policies. One challenge is that local governments cannot be trusted to make socially beneficial decisions about land use and development in hazardous areas. A second, related challenge is that there is an uneven playing field in making sustainable management decisions so that development decisions will not be fair.

Our data about perceived changes in development decision-making in New South Wales and in New Zealand show a strong movement toward more flexibility in rules and greater willingness of local officials to negotiate development decisions. This is consistent with the findings of the preceding chapter. This is also consistent with the notion of case-by-case decisions on the merits of development that is an explicit goal of the New South Wales policy, and is consistent with a shift under the *Resource Management Act* from prescriptive rules to consideration of effects in seeking sustainable environmental management.

There is only partial support for the concern that cooperative policies open avenues for the development community to capture local decision-making about development in hazard-prone areas, thereby leading to increased exposure. This potential is greatest among those local governments with increased flexibility in rules, for which developer influence is strong, and for which development demand is strong. We observed this in our case study of Lismore. The city has had steady growth pressure, and its elected representatives, who are responsive to the interests of businesses from the flood-prone central business district, are generally pro-development.

Other cases, and our analysis of perceived effectiveness, suggest there is less likelihood of increased exposure among those places that are serious about hazard management and that are willing to negotiate acceptable outcomes with developers. Respondents appear to believe that they can use negotiation as an effective means for obtaining outcomes that do not increase exposure. Being serious about hazard management presumably means that local officials are more effective in managing development in hazardous areas, thereby decreasing, or at least preventing increases in, exposure. We observed this for our case study of Fairfield where there was a strong commitment to dealing with flood hazards.

The cases we have discussed illustrate a range of possible outcomes under cooperative intergovernmental policies. These relate both to the responses of different localities in managing flood hazards and to the outcomes of those efforts in reducing exposure while also increasing community sustainability. The positive points are that cooperative policies seem to allow the necessary flexibility in local decision-making to permit choices that foster sustainability, as found in the case discussion of Fairfield. Yet, the cooperative policies also leave the door open for non-compliance, as found in Liverpool, or for choices that do not promote long-term sustainability, as seems to be the situation in Lismore. Although a variety of factors contribute to these differential outcomes, the commitment of elected officials and staff to addressing hazards looms large in the overall equation.

NOTES

1 See, for example: National Commission on the Environment, *Choosing a Sustainable Future* (Washington, D.C.: Island Press, 1993), pp. 113–117.

2 For a review of the literature surrounding these challenges and policy responses within the American setting see Chapter 1 of Raymond J. Burby and Peter J. May with Philip R. Berke, Linda C. Dalton, Steven P. French, and Edward J. Kaiser, *Making Governments Plan: State Experiments in Managing Land Use* (Baltimore: Johns Hopkins University Press, 1996).

3 The issue of tradeoffs in risk reduction has been advanced by those writing about the economics of risk. Classic treatments of this issue are found in Douglas C. Dacy and

Howard Kunreuther, *The Economics of Natural Disasters* (New York: Free Press, 1969); and Clifford S. Russell, "Losses from Natural Disasters," *Land Economics* 46, no. 4 (November 1970): 383–393.

4 See: Ian Burton, Robert W. Kates, and Gilbert F. White, *The Environment as Hazard*, 2nd edition (New York: Guilford Press, 1993), pp. 262–263.

5 The percentages of respondents reporting changes in exposure are greater for those councils in New South Wales that have substantially completed the merits planning process than those who have not participated in the process. For those substantially completing the process, the percentage reporting increases is 20 and the percentage reporting decreases is 24. For those not undertaking the process, the corresponding percentages are 8 and 16.

6 The survey of local governments in Florida did not include questions that are directly comparable to those concerning perceptions of change in the other surveys. The two questions reported here asked about the role of state government and policy in bringing about change and the benefits that respondents attach to the plans created under the Florida policy.

7 The most sophisticated analysis of this type that we are aware of is a study of the United States' National Flood Insurance Programs impacts in affecting development in flood-prone areas. See: Raymond J. Burby, Scott A. Bollens, James M. Holloway, Edward J. Kaiser, David Mullan, and John R. Sheaffer, *Cities Under Water: A Comparative Evaluation of Ten Cities' Efforts to Manage Floodplain Land Use*, Program on Environment and Behavior Monograph No. 47 (Boulder, CO: Institute of Behavioral Science, University of Colorado, 1988.)

8 The initial databases of buildings and the re-surveys undertaken in 1993 and 1994 are based on the methodology used for a computer damage assessment program, ANU-FLOOD. Technical descriptions of the ANUFLOOD methodology are available in Mark A. Greenaway and D. Ingle Smith, *ANUFLOOD: Programmer's Guide and User's Manual* (Canberra: Centre for Resource and Environmental Studies, 1983), and D. Ingle Smith and Mark A. Greenaway, *ANUFLOOD: Field Guide* (Canberra: Centre for Resource and Environmental Studies, 1983). General application of the ANUFLOOD program is illustrated in D. Ingle Smith and Mark A. Greenaway, "The Computer Assessment of Urban Flood Damages: ANUFLOOD" in *Desktop Planning, Advanced Microcomputer Applications for Physical and Social Infrastructure Planning* ed. P. W. Newton, R. Sharpe, and M. A. P. Taylor (Melbourne: Hargreen Press, 1988), pp. 230–250.

9 The levee is not "tied," thereby allowing flooding behind the structure in the event of a major flood. Because of this design, a major flood would have markedly reduced flood velocities, thereby reducing the risk of damage from building failure. Resilience to such flooding is further increased by a floor-height regulation that requires habitable floors to be at least 0.3 meters above the 100-year flood level.

10 The term probable maximum flood (PMF) is used here to indicate the theoretically worst flood that could occur. Methods for its estimation are a matter of dispute, but various techniques are available. Consideration of the PMF is implicit in the hydrological studies recommended as part of the merits planning process. The degree to which this has been achieved among local governments in New South Wales, following the introduction of the state's merits flood policy, is uneven.

11 The newer design of levee systems is intended to reduce the hazard posed by high-velocity flood waters, and therefore reduce the chance of failure. Levees like that of Singleton and Muswellbrook are "untied" so that flood water comes around the end. As a consequence, the "protected" areas are subject to backwater flooding, but not sudden increases in the depth of inundation from levee overtopping or failure.

APPENDIX

Table 9.A Explaining variation in perceived outcomes (New South Wales)

Explanatory variables	*Logistic regression models explaining perceptions of change in:*[a]		
	Council willingness to negotiate	Fairness in decision-making	Exposure to natural hazards
Development controls:			
Council effort to manage development	−.48	.60**	−.56
Flexibility in rules governing development	1.26***	.92***	2.87**
Willingness to negotiate with developers	—	.83**	−2.49**
Political factors:			
Influence of developers	.15	−.52*	2.71**
Influence of neighborhood groups	.31	−.01	−.11
Council capabilities:			
Staff professionalism	.86**	−.33	.26
Staff capacity	−.02	.54**	−1.41**
Development situation:			
Demand for development in hazardous areas	.50	−.20	3.17**
Amount of hazard-free land	−.59*	−.02	−1.93**
Growth in population	.52	1.12	8.21**
Population	.31	−.31	−7.12**
Degree of risk:			
Prior losses	.44	.46*	.89
Risk from flooding	−.42	.23	—
Constant	−.22	−.50	−1.01
Number of cases	84	84	39[b]
Model statistics:[c]			
Percentage correctly classified	83	79	85
Model chi-square significance	<.01	<.01	.02
Pseudo-R^2	.38	.32	.53

Notes:
  * $p < .10$** $p < .05$*** $p < .01$ (one-tailed tests)

continued

[a] Cell entries are logit coefficients from logistic modeling involving a comparison for each dependent variable of councils reporting increases with those reporting decreases or no change in the item (exposure model excluded responses of no change). Positive signs for coefficients indicate the factor contributes to a greater likelihood of the dependent variable being increased. Negative signs for coefficients indicate a reduced likelihood of the dependent variable being increased. Because explanatory factors were standardized prior to model estimation, the magnitude of the logit coefficient serves as a general indicator of the relative strength of effect of a factor.

[b] The number of cases is lower than the other dependent variables because those councils for which perceived exposure did not change are excluded for this model.

[c] Percentage correctly classified measures the percentage of observations that are correctly predicted for the model. The model chi-square significance measures the goodness-of-fit of the overall model for which values less than .05 are generally accepted as good fits. The McFadden pseudo-R^2 is an analog to the regression coefficient of variation.

Part III

THE POLICY
INNOVATIONS REVISITED

INTRODUCTION

The assessment of the policy innovations provided in Part II provides the basis for summarizing the strengths and limitations of approaches that national or state governments can use to persuade local governments to be good stewards of the environment. The central challenge for either the coercive or the cooperative approach is to build the commitment of local officials to higher-level policy goals. Other challenges are to reduce gaps in the compliance of local governments with the procedural or substantive requirements of higher-level mandates, and to better tailor higher-level policies to accommodate differing local circumstances. These latter challenges are present in different degrees for each approach examined in this book to the governance of environmental management.

Part III examines responses to the challenges presented by the different approaches to environmental management and presents the implications of our research for the design of environmental policies. Chapter 10 addresses the "commitment conundrum" that stems from the indifference of a substantial minority of local government officials to environmental sustainability. Chapter 11 summarizes our findings concerning the strengths and limitations of the cooperative and coercive approaches. We conclude with attention to the implications of our research for the governance of environmental management.

10

THE COMMITMENT
CONUNDRUM

The preceding chapters document variability in the efforts that local governments have undertaken to either manage development in hazard-prone areas or otherwise address risks posed by natural hazards. A key aspect of that variability is the low levels of commitment of a substantial subset of local-government officials to those undertakings. This is a serious problem that results in half-hearted efforts and, in some instances, outright failure to comply with higher-level mandates. In either case, lack of such commitment serves as a key obstacle to achieving sustainability with respect to natural hazards. Gaps in commitment among local governments to hazard-mitigation goals also point to limitations of the intergovernmental mandates that seek to enhance attention by local governments to these problems.

The idea that commitment is an important factor in policy implementation has been an enduring tenet of the public policy literature. In studies of efforts by local governments to reduce natural hazards, the lack of commitment is frequently pointed to as a key explanation for poor local government performance. Researchers in the United States in the mid-1970s surveyed 2,000 leaders in state and local government and the private sector to find out how concerned they were about a variety of natural hazards. The researchers concluded: "For the most part, political decision makers in the states and local communities do not see environmental hazards as a very serious problem, particularly in comparison with many other problems these governmental units are expected to be doing something about."[1] Other researchers have found direct links, similar to those we reported earlier in this book, between low commitment and local government failure to address flood hazards, coastal storm and hurricane hazards, and seismic safety. Findings such as these led analysts William Petak and Arthur Atkisson to conclude, "The primary impediment to the adoption and enforcement of effective natural hazards regulatory policy has to do with the 'willingness' rather than the 'capacity' of governmental law-making bodies to act . . ."[2]

This chapter addresses the variability in commitment of local governments to hazard management and, by extension, the variability in commitment to environmental management. We seek to understand why commitment is so variable and to identify what higher-level governments can do to solve the commitment

196

conundrum. Commitment is not a black box. Higher-level governments have a number of means they can use to build commitment of lower-level governments and with it a better record in achieving broad state and national policy goals.

COMMITMENT UNDER COERCIVE AND COOPERATIVE MANDATES

Our surveys of local governments in Florida, New South Wales, and New Zealand reveal similar failures in the willingness of local government to address natural hazards. In each setting, we found a substantial minority of localities that have little commitment, particularly among elected officials, to the hazard-mitigation goals of higher-level mandates. A quarter of the localities in Florida, two-fifths of those in New South Wales, and a fifth in New Zealand are served by elected officials whom their staff consider are only lukewarm about hazard mitigation. That is, they have either no or low commitment to reducing the threat of losses from natural hazards. The proportion of governments with uninterested staff is much lower (between a tenth and a twentieth of the governments in each setting), reflecting closer contact between staff and higher-level government officials and possibly the more direct responsibility of staff for dealing with hazards.

Information that is available about changes in commitment before and after the planning mandates that we study were established indicates that commitment is increasing over time in each setting. In New South Wales, about a third of the councils we surveyed reported that the commitment of elected officials to hazard mitigation had increased since the state instituted the merits flood policy, and over half reported more commitment from planning staff. Only a few respondents in New South Wales reported a decrease in commitment. In New Zealand, the results were similar. A third reported increases in the commitment of elected officials since passage of the *Resource Management Act* and over half reported that planning staff were now more committed to reducing natural hazards. In Florida, we did not ask a similar question, but we have data on the priority that sixteen of the thirty local governments surveyed in 1990 gave to hazard mitigation in 1979, six years before the 1985 growth management legislation was enacted.[3] This comparison indicates that commitment has increased as evidenced by the fact that respondents from 74 percent of the jurisdictions reported a high degree of commitment in 1990 versus 57 percent reporting similar levels in 1979. The change in commitment in Florida is similar in magnitude to the changes reported in New South Wales and in New Zealand.

Intergovernmental policy influence on commitment

The New South Wales data provide a more refined, albeit more speculative, understanding of the underlying mechanisms for the impacts that cooperative policies have on the commitment of local governments. Our survey respondents' rating of the impact of the state's merits flood policy on the commitment of staff is much greater than the corresponding impact on elected officials. Seventy percent of respondents from local councils participating in the state's floodplain-planning process reported increased staff commitment after introduction of the merits flood policy, whereas only 42 percent reported increases in the commitment of elected officials (less than a handful reported decreases).

This differential impact among staff and elected officials suggests that co-operative policies primarily influence staff commitment. That, in turn, provides a basis for advocating solutions for adoption by elected officials. Our statistical analysis of changes in commitment helps bear this hunch out. The amount of technical and other forms of assistance provided to local governments by state agencies under the merits policy is a noteworthy predictor for changes in commitment among local government staff, but it is not a noteworthy predictor of changes in commitment among elected officials.[4] By working with local governments that choose to participate in a planning process to help solve devel-opment management problems, state administrations can foster commitment among staff members in those governments that come to recognize the extent of problems and the available solutions.

Our guess is that the situation is different for coercive policies, as illustrated by the experience in Florida. The fear of sanctions for not participating drives procedural compliance of local governments with state requirements, as observed in Chapter 7, and develops a calculated commitment among elected officials. This is then augmented by the effects of carrying out planning processes that foster increased commitment by staff as they learn about the problem, and as citizens lobby for solutions.

How these differential impacts play out over time is unclear. Assuming strong monitoring and enforcement, coercive mandates are presumed to have an edge in building commitment from local government because they directly impact on calculations by local officials about whether to comply. This, however, is a short-run argument. The promise of cooperative policies is that in allowing greater flexibility in local actions they can sustain and enhance commitment over time. Commitment may erode under coercive mandates over time, particularly as monitoring and enforcement ease. But, this will not necessarily happen, especially if community participation in planning processes under the coercive mandate lead to the types of outcomes discussed in what follows.

KEYS TO THE PUZZLE OF LIMITED COMMITMENT

The preceding discussion suggests that both coercive and cooperative inter-governmental policies influence the commitment of local governmental officials. But, the influence of each is constrained. Coercive approaches build calculated commitment as long as strong oversight is provided. Cooperative approaches enhance normative commitment but only among those localities willing to engage in the collaborative-planning process. The Catch-22 for cooperative regimes is that they require an initial level of commitment to engage that process. This was missing for a substantial number of jurisdictions in New South Wales. Given this conundrum and the limitations of intergovernmental mandates in securing greater commitment, what can higher-level governments do? Understanding the problem is key to solving the puzzle. In this section, we consider different aspects of the problem, and then examine the extent to which each plays a role in limiting the commitment of local governments.

Barriers to commitment

It is surprising that the basis for the lack of commitment has not been more systematically studied. We think psychological, political, and practical considerations hold the keys to this puzzle.

Psychological constraints The policies in each setting ask local officials to create or follow new ways of managing land use and development. These tasks present a psychological hurdle. Scholars of local government decision-making suggest that because community leaders develop stable patterns of policy-making, it is difficult to counteract the prevailing momentum.[5] Indeed, the need to overcome inertia in local policy-making provides one justification for mandates from higher-level governments. The occurrence of a natural disaster can shake local policy-makers loose from their established policy routines, but disasters are an imperfect policy tool. They occur infrequently and often produce short-sighted responses to the problem as shown by the case of Lismore in Chapter 8.

A related psychological barrier stems from the nature of "public risks" that are " . . . broadly distributed, often temporally remote, and largely outside the individual risk bearer's direct understanding and control."[6] Policy-makers, like other individuals, are motivated by rewards and the likelihood of attaining them. They see few rewards from changing past policy-making patterns and devoting resources to natural hazards. Policy-makers, like most other people, are not particularly inclined to worry (or even think) about low-probability events.[7] If policy-makers become aware of and actually do think about hazards, the perceived remoteness in time of low-probability hazards (e.g., a 100-year flood) leads to sharp discounting of the benefits of avoided costs. Also, policy-makers may become fatalistic when thinking about some hazards, such as earthquakes, which they can neither prevent from occurring nor even estimate very accurately where if not when losses are likely to occur. These flaws in hazard perception suggest that rather than waiting for a disaster to occur, governments may be able to change commitment by making local officials more aware of the risks posed by natural hazards.

Political barriers Psychological barriers are not limited to policy-makers. They also affect their constituents, who are likely to see little need to demand governmental attention to hazards and little benefit from changing their behavior in response to building and land-use regulations aimed at reducing risks. Without positive signals from their constituents, politicians who do not see the problem are unlikely to devote resources to hazard mitigation. If the signals they receive are all negative, local policy-makers may actively resist higher-level mandates to regulate private activities, such as new development, in hazard-prone areas. For example, in the United States, active opposition by real estate and development interests has been a factor limiting attention by local governments to hazard mitigation.[8]

This is not to say that the potential for political support for policies to reduce risks is always absent. Particularly after a disaster, groups who have suffered losses are likely to lobby for loss-reduction measures. However, the measures they favor tend to be those that limit the event (e.g., flood control and shoreline protection works) rather than those that manage development and exposure to it. Increasing the awareness that various constituents have of threats posed by hazardous events may go a long way toward persuading policy-makers to support measures for reducing natural hazards. But, awareness alone may be insufficient if the goal is sustainable development rather than solely risk reduction. As illustrated in the previous chapter, flood-control measures can foster unsustainable urban development in areas at risk.

The necessary constituency building is likely to require a collaborative, participatory planning process in which a key goal is social learning.[9] That is, the various stakeholders in hazard mitigation need to both be informed of the risks and convinced that environmentally sustainable solutions to the problem are reasonable for all concerned interests. If some degree of consensus or policy convergence can be attained, the political support needed to sustain local commitment to policy goals may be attained.

Practical barriers Psychological and political obstacles that stand in the way of commitment appear to be formidable. Less serious, at least in the sense that they are more tractable, may be practical barriers to commitment. These barriers include the complexity of policies to achieve sustainability and the lack of local expertise and financial capacity to either understand or implement them. If one accepts the proposition that it is difficult to become committed to courses of action for which there is little ability to pursue, then a lack of local capacity to pursue hazard-mitigation policies may also account for shortfalls in commitment.

To the extent that is true, commitment might be enhanced if mandates include capacity-building elements, such as financial and technical assistance. As discussed in Chapter 1, capacity-building is a central feature of cooperative intergovernmental mandates. The preceding discussion suggests it could be critical to the success of cooperative mandates, since they lack coercive elements to force compliance when commitment is lacking.

Plate 10.1 Exposed lighthouse in Cape Saint George, Florida – A monument to the psychological and political barriers to addressing hazards (courtesy of Phil Flood, Florida Department of Environmental Protection)

Examining variation in commitment

Each of the three barriers to commitment – psychological, political, and practical – may help explain the gaps in procedural compliance by local governments that we documented in Chapter 7 and the variation in hazard-management practices that we discussed in Chapter 8. In order to more systematically examine the variation in commitment, we undertook statistical analyses that addressed each of the barriers. Initially, we found that these factors did little to explain variation in commitment of local governments in New Zealand. We believe that is due to disruption caused by local government reorganization in the late 1980s, and the turmoil caused by new processes for environmental management. Given the limitations of the New Zealand data, we focus on Florida and New South Wales in the remainder of this chapter.

In separate statistical analyses of the commitment of local governmental elected officials and of their staff, we found that political demands by groups to address natural hazards and actually experiencing a serious disaster have strong impacts on the degree of commitment.[10] Experience with catastrophic losses is the primary way elected officials become committed to hazard mitigation and an important factor in explaining staff commitment. Constituent demands that action be taken to address hazards also have a noteworthy impact. These demands can come from a variety of groups, including local businesses, environmental groups, and neighborhood organizations. We also found that commitment of elected officials for Florida's local governments, with its coercive state mandate, is greater on average (when controlling for other influences) than that of elected officials in New South Wales.

A number of factors that relate to psychological and practical barriers seem to be less important. Suffering a catastrophic disaster is a harsh way to learn about the importance of hazard mitigation, but neither experience with chronic losses nor disaster potential affect commitment of local officials or staff in Florida and New South Wales. The capacity of local governments to address hazards also has little effect on commitment. Possibly, once some minimal level of staff competence to deal with natural hazards is achieved, commitment is no longer sensitive to additional increments of capacity. However, we could not detect that threshold with the data on hand.

Although commitment among elected officials tends to be lower in New South Wales than in Florida or New Zealand, the relevant state agencies of New South Wales have attempted to enhance normative commitment through extensive technical assistance in implementing the merits approach to floodplain management. These measures have begun to pay off. Where regional offices have done more to implement the mandate and deliver technical assistance, local government commitment is higher.[11]

In summary, we have some evidence, though it is inconclusive, that the commitment-building features of a coercive mandate, such as Florida's growth management legislation, and the degree of effort expended in implementing a cooperative mandate, such as the merits policy in New South Wales, can affect the commitment of local officials to higher-level goals. Important elements in building such commitment are the demands that political constituencies place on governments for attention to natural hazards and the actual occurrence of a disaster.

ADDRESSING THE COMMITMENT CONUNDRUM

If constituency demands have an important role in persuading local officials to pay attention to natural hazards, what can national or state governments do to build such demand? Three options seem plausible. One is to refrain from interfering in local political processes and wait for a disaster to serve as a catalyst for constituency demands. A second is to attempt to foster constituency demands by providing information about risks posed by natural hazards and sustainable ways of living with them. A third, highly proactive approach, is to use participatory-planning processes to create a consensus on the need for governmental action. The data we collected in Florida and New South Wales allow us to comment on the efficacy of each approach for building constituency demands.

Doing nothing

Doing nothing to influence constituency demands has some logic. Like local officials, constituencies for hazard mitigation respond to the nature of the problem. As losses are experienced and the threat of loss increases, there can be increased demands by various groups for attention to the problem. This was illustrated in the previous chapter by the role of major floods in drawing community attention to flood hazards in Fairfield, New South Wales. Table 10.1 compares constituency demands among local governments in Florida and New South Wales, while taking into account different levels of prior losses from natural disasters and potential exposure to flood hazards. Constituency demands are measured by an index of various actions (sought information, asked for community action, attended meetings, or served on relevant committees addressing the problem) by local business groups, environmental groups, neighborhood groups, and individuals not associated with any particular group. Higher scores indicate greater demands. In Florida, both prior losses and the potential for loss are associated with demands for governmental action to

Table 10.1 Risks and constituency demands to address them

Indicator and categories	Mean of constituency demands for local governments in:[a]	
	Florida	New South Wales
Catastrophic losses[b]		
None or small losses	5.1	3.1
Moderate or greater losses	12.0	5.8
Difference (above categories)	6.9**	2.7***
Potential losses[c]		
Low	3.0	3.1
Moderate	5.3	5.7
High	7.7	4.7
Difference (high vs. low)	4.7	1.6***

Sources: Compiled by authors from surveys of local governments
Notes:
*   p <.10** p <.05*** p <.01
[a]   Cell entries are means of constituency demand scores for the specified indicators and categories. Higher scores indicate greater demands from various groups (theoretical maximum is sixteen). Cell entries for the "difference" rows are the difference of means between the categories noted in parentheses. Statistical tests are tests for difference of means. Based on responses from thirty local governments in Florida and 127 local councils in New South Wales.

[b]   Florida: whether since 1970 the jurisdiction was ever declared for federal disaster aid as a result of a list of natural disasters; and if so, the extent of damage. New South Wales: whether since 1973 there had been more than incidental damage to property or loss of life from flood hazards; and if so, the extent of damage.

[c]   Respondent rating of the demand for new development in hazard-prone areas of the jurisdiction.

reduce risk. In New South Wales, potential losses have less effect, possibly because development pressures in hazardous areas are markedly lower.

Since constituency demands generally reflect the severity of the problem, national or state governments could conclude that their intervention is not essential. From this perspective, local constituency demands may be adequate for building commitment where it is most needed – in the most hazard-prone locales. Even though we do not analyze data for New Zealand, it is worth noting that the central government has the express aim of moving responsibility for disaster losses onto individuals and local communities in the hope that they will act prudently. The cost of doing nothing, however, is slippage in applying regulatory measures that prevent hazards from becoming a problem in the first place.

This presents a policy paradox. On the one hand, preventive planning measures work best before hazardous areas develop. Once development gets started and demand builds, land-use and building-regulatory measures are not very effective in preventing losses. On the other hand, before development takes place and a problem is created, there is little interest among local governments in enacting hazard-prevention measures. As a consequence, doing nothing runs the risk that over time a serious problem will develop that it is too late to adequately address.

Awareness-building strategies

If one subscribes to the view that public managers work at the seams of government in shaping the political environment as well as in responding to it, then it is appropriate to consider ways to build local constituencies for managing the environment. Providing citizens and interest groups with information about risks and appropriate adjustments is a non-intrusive approach that many believe will help create political constituencies. For example, William Anderson and Shirley Mattingly argue that:

> The public's perception of risk strongly affects the emergency manager's work. An informed public can be a major ally: awareness can lead to action, including pressure on legislators and other policy-makers. . . . In short an informed and concerned public wields a significant amount of power and a wise emergency manager will ensure that the public stays informed so that public power – and public policy – contribute to the strength of the emergency management program.[12]

Information can help to transform an objective problem that is inimical to society's welfare into one, that the public perceives as a problem. It is the perceived problem, not the objective one that is critical for public policy formulation.

Information strategies have several possible drawbacks. They can be costly and whether they work to increase public attention to hazards is unclear. A comprehensive review of public information campaigns by Janet Weiss and

Plate 10.2 Disaster awareness efforts in Australia (courtesy of the Disaster Awareness Program, Emergency Management Australia)

Mary Tschirhart provides the caution that "[T]he conventional wisdom about public information campaigns . . . [is] that they are trivial or ineffectual policy instruments."[13] Here, however, we are interested in information which can build public understanding of a policy problem and ways of dealing with it rather than the behavioral change (e.g., turning off lights to save energy) sought by information programs that have been found to have little impact.

Because information is not value neutral, it can be controversial as well. As noted in Chapter 4, the attempt in New South Wales to make floodplain maps widely available created considerable controversy because landowners thought the information would devalue their property. When information campaigns are used consciously to manipulate public opinion and manage local political processes, they can be viewed as subverting democratic values.[14] Sometimes the distinction between information and politically inspired propaganda is subtle. If information programs are used to foster a more informed electorate and wiser decision-making by individuals, with enhanced demands for government attention to reducing natural hazards as a secondary benefit, they seem unlikely to run into claims that government officials have overstepped the bounds of propriety.

Information programs have been employed in both Florida and New South Wales, although in both cases the goal has been to persuade households to take self-protective behavior rather than to mobilize demands by interest groups.[15]

205

In Florida, information programs include flood and hurricane-hazard maps provided by local governments as part of the National Flood Insurance Program, and regional storm-surge maps provided by regional councils and counties as part of emergency preparedness plans. The Community Rating System of the National Flood Insurance Program, which provides reduced flood-insurance rates in exchange for community actions to reduce risk, provides an incentive for communities to mount public information campaigns and enact laws that require real estate agents to disclose hazards prior to the sale or rental of property. In New South Wales, information campaigns include distribution of flood-awareness brochures to households, media liaison, detailed technical reports, and distribution of information about flood-warning systems.

Of the local governments we studied in Florida, 90 percent have put in place programs to increase awareness of risks posed by natural hazards. In addition to maps showing areas at risk, a majority of the local governments in Florida have mounted campaigns to educate the public about the risks. Less frequently used strategies include promotion of flood insurance (employed by 37 percent), voluntary disclosure of hazards prior to sale or rental (encouraged by 13 percent), and mandatory disclosure of hazards (3 percent). In addition, 40 percent of the local governments have warning systems in place that should stimulate awareness of hazards. On a seven-point scale of effort devoted to awareness-building, the average local government in Florida scores at the mid-point. We do not have similar information about specific information strategies in New South Wales, but governments there on average devote less effort to awareness-building activities than in Florida – possibly because of the mapping fiasco discussed in Chapter 4.[16]

Given this variation in awareness-building efforts, we can examine the possible effects of a more informed public (or subsets of it) on local political demands to address hazards. We recognize that government efforts to provide information may, themselves, be a product of constituency demands for attention to natural hazards. Because we lack adequate before-and-after data, it is impossible to test conclusively which came first, information or constituency demands. While recognizing the potential for reverse causation, we think some degree of information (including actually experiencing the consequences of natural events) stimulates citizens to demand government attention to the problem, which in turn leads government to develop and disseminate more information.

Table 10.2 shows that the more effort governments devote to informing the public about hazards, the more various stakeholders become active in demanding attention to loss reduction. As with the preceding analysis, these stakeholders include local business groups, environmental groups, neighborhood groups, and individuals not associated with any particular group. In both Florida and New South Wales, constituency demands are twice as high in localities that expend a high degree of effort on information programs compared to those that expend little effort on these activities. The relationship remains strong when a variety of other factors that can stimulate political action are controlled statistically. These

Table 10.2 Public awareness-building efforts and constituency demands

Efforts to build public awareness[b]	Mean of constituency demands for local governments in:[a]	
	Florida	New South Wales
Low	3.8	3.1
Moderate	5.3	5.0
High	7.5	6.0
Difference (high vs. low)	3.7	2.9***

Sources: Compiled by authors from surveys of local governments
Notes:
  * p <.10 ** p <.05 *** p <.01
  [a] Cell entries are means of constituency demand scores for the specified categories of local governmental efforts to build awareness. Higher scores indicate greater demands from various groups (theoretical maximum is sixteen). Cell entries for the "difference" row are the difference of means between the high and low categories of efforts to build awareness. Statistical tests are t-tests for difference of means. Based on responses from thirty local governments in Florida and 127 local councils in New South Wales.

  [b] Respondent rating of the degree of effort the jurisdiction devotes to measures to increase people's awareness of hazards.

data suggest public information can be a tool for building a constituency for hazard mitigation, which in turn can help build commitment among local officials to do something about the hazards problem.

Planning strategies

Planning strategies take information one step further by actually working with potential constituency groups to build understanding of the problem and find acceptable solutions.[17] As discussed in Chapter 7, central components of the hazard and environmental polices examined in this book are requirements for local planning and citizen participation processes. Florida's growth management legislation and the administrative rules implementing it specifically demand that local governments involve citizens in planning, and citizens are authorized to challenge plans before the state if those requirements are ignored. In New South Wales, the *Environmental Planning and Assessment Act* mandates citizen participation in the preparation of environmental plans. The merits flood policy puts formation of a floodplain management committee, which is broadly representative of stakeholders, as the first step in the planning process. New Zealand's *Resource Management Act* also stipulates extensive public consultation and awareness processes.

Planning can empower citizens to become active in demanding government attention to hazards in two ways. First, the information provided in a plan (the plan's "fact-basis"), like the information strategies discussed above, serves a "frame-setting" function that helps both government decision-makers and constituency groups better understand their situation.[18] Second, where plans actively involve citizens in problem solving, they can reduce conflict and produce consensus about how government should address hazards.[19] By fostering what planning theorists term "communicative rationality" rather than (or in addition to) "technical rationality", plans based on participation can result in broad-based support for the recommendations contained in plans and, in turn, greater commitment from policy-makers to address the relevant problems.

We examined the extent to which aspects of planning processes have promoted constituency action to secure government attention to hazards. For Florida, where each of the local governments we studied had adopted a plan in response to the state planning mandate, we have information about the quality of local plans.[20] Higher quality plans with better factual underpinnings, more goals related to hazards, and more policy recommendations should be accompanied by greater demands for governmental action from citizens. In New South Wales, less than half of the governments we studied have prepared floodplain management plans and just over half have formed the floodplain management committees the state recommends. Rather than looking at plan quality for these governments, which would provide a very selective understanding, we assess whether demands for governmental attention to flood hazards are greater where the local plans and committees exist than where they do not exist.

The relationship between planning and constituency demands is depicted in Table 10.3. These data show that in both Florida and New South Wales constituency demands for local governments' attention to natural hazards are higher where plans exist and citizen committees have been formed. The difference in plan quality in Florida, while positive, is not statistically significant, due in part to the small number of local governments in our Florida sample. Subsequent analyses revealed that the differences in New South Wales wash out when other factors that can stimulate constituency demands (e.g., previous losses and the state government's awareness-building efforts) are controlled statistically. It may be that the more technical floodplain-management planning of the merits flood policy is less successful in building a community-based constituency than are broader-based planning processes such as those found in Florida and New Zealand.

CONCLUSIONS

In this chapter, we have explored what we term the "commitment conundrum" that arises when local governments are indifferent to sustainability. This indifference is evident for substantial minorities of the governments we studied, with about a third having elected officials with little or no interest in hazard mitigation – the aspect of environmental sustainability examined in this book. That is a serious problem since, as we have shown in earlier chapters, commitment is a lynch pin in placing hazard mitigation (and sustainability) on local political agendas and in providing the impetus for vigorously pursuing implementation once relevant policies are adopted.

Table 10.3 Local government planning and constituency demands

Indicator and categories	Mean of constituency demands for local governments in:[a]	
	Florida	New South Wales
Hazard mitigation plan[b]		
Low quality / None	4.9	3.4
High quality /Plan completed	6.1	5.2
Difference	1.2	1.8 ***
Floodplain management committee[c]		
Not established	–	3.3
Committee formed	–	5.4
Difference	–	2.1 ***

Sources: Compiled by authors from surveys of local governments

Notes:

* p <.10 ** p <.05 *** p <.01

[a] Cell entries are means of constituency demand scores for the specified indicators and categories. Higher scores indicate greater demands from various groups (theoretical maximum is sixteen). Cell entries for the "difference" rows are the difference of means between the categories. Statistical tests are t-tests for difference of means. Based on responses from thirty local governments in Florida and 127 local councils in New South Wales.

[b] Florida: Index of plan quality for which those above the median score for all plans are considered high quality, and below the median low quality. New South Wales: whether the jurisdiction has completed a floodplain management plan or not.

[c] Whether a floodplain management committee has been formed or not; only applies to jurisdictions in New South Wales.

Three conditions can contribute to limited commitment: psychological barriers to understanding the gravity of natural hazards; political indifference or outright resistance to policy change among constituency groups or elected officials; and lack of capacity to deal with complex hazard-mitigation strategies. Of these three, we found evidence that the first two affect how Florida's growth management program and New South Wales' merits flood policy have played out. (The New Zealand policy experience was too new to analyze for these purposes.) Commitment to the goals of the mandates we studied tended to be lower when risk is not self-evident and when various interest groups make few demands for governmental attention to the problem. The capacity of local governments to carry out hazard-mitigation programs, within the range of variation that we observed, does not seem to constrain commitment.

We have considered three strategies – doing nothing, building awareness of hazards, and undertaking participatory planning – for addressing the commitment conundrum. Doing nothing makes some sense since local constituency demands may be adequate for building commitment where it is needed; that is, in the most hazard-prone locales. However, this strategy runs the risk that over time a serious problem will develop that is then too late to adequately address. In

Florida and New South Wales, informational strategies are stimulating demands from constituents for attention to natural hazards. Engaging citizens and groups in participatory-planning processes for addressing natural hazards can also have positive effects in creating constituencies. However, our findings show that the form of the participatory process makes a difference.

The bottom line is this. The commitment conundrum can be dealt with through adequate policy design and strong implementation. When the commitment of local governments to higher-level policy objectives is likely to be weak because of inadequate appreciation of the policy problem, information programs and participative planning provide useful tools for building the commitment of local elected officials. This is a critical ingredient if local governments are to become willing partners in pursuing state and national policy goals. This chapter suggests both informational approaches and planning processes are desirable ingredients for intergovernmental mandates. They can build the basis of understanding and support that is needed to sustain commitment to higher-level policy goals.

NOTES

1 Peter H. Rossi, James D. Wright, and Eleanor Weber-Burdin, *Natural Hazards and Public Choice: The State and Local Politics of Hazard Mitigation* (New York: Academic Press, 1982), p. 9. Hazards ranked just behind pornographic literature as a policy problem.

2 William J. Petak and Arthur A. Atkisson, *Natural Hazard Risk Assessment and Public Policy* (New York: Springer-Verlag, 1982), p. 422. For research documenting lack of local government commitment with respect to various natural hazards see: Philip R. Berke and Timothy Beatley, *Planning for Earthquakes: Risk, Politics and Policy* (Baltimore, MD: Johns Hopkins University Press, 1992); Raymond J. Burby and Steven P. French with Beverly A. Cigler, Edward J. Kaiser, David H. Moreau, and Bruce Stiftel, *Floodplain Land Use Management: A National Assessment* (Boulder, CO: Westview Press, 1985); and David R. Godschalk, David J. Brower, and Timothy Beatley, *Catastrophic Coastal Storms: Hazard Mitigation and Development Management* (Durham, NC: Duke University Press, 1989).

3 Results of the 1979 survey are summarized in Raymond J. Burby and Steven P. French *et al.*, *Flood Plain Land Use Management*, op. cit. Also see: Raymond J. Burby and Steven P. French, "Coping with Floods: The Land Use Management Paradox," *Journal of the American Planning Association* 47, no. 3 (Summer 1982): 289–300.

4 The assistance measure is computed for each jurisdiction as the sum of six possible state agency actions (each 1 = yes, 0 = no): respond to questions, reviewed plan or policy, distributed guidance materials, distributed example plan or policy, telephone consultation, on-site technical assistance. Logistic regression analyses of changes in commitment, controlling for the same set of explanatory variables of regressions reported later in the chapter, suggest greater impacts of assistance on commitment of staff (logit = .66, p <.05) than the influence on the commitment of elected officials (logit = .14, p = .33).

5 See, for example: Heinz Eulau and Kenneth Prewitt, *Labyrinths of Democracy: Adaptations, Linkages, Representation, and Policies in Urban Politics* (Indianapolis: Bobbs-Merrill, 1973); and Ira Sharkansky, *The Routines of Politics* (New York: Van Nostrand Reinhold, 1970).

6 Peter W. Huber, "The Bhopalization of American Tort Law," in National Academy of Engineering, *Hazards: Technology and Fairness* (Washington, D.C.: National Academy Press, 1986), p. 90. Also see: Peter J. May, "Addressing Public Risks: Federal Earthquake Policy Design," *Journal of Policy Analysis and Management* 10, no. 2 (1991): 263–285.

7 See Colin F. Camerer and Howard Kunreuther, "Decision Processes for Low Probability Risks: Policy Implications," *Journal of Policy Analysis and Management* 8, no. 4 (Fall 1989): 565–592; and Paul Slovic, Baruch Fischhoff, and Sarah Lichtenstein, "Behavioral Decision Theory Perspectives on Protective Behavior," in *Taking Care: Understanding and Encouraging Self-Protective Behavior,* ed. Neil D. Weinstein (Cambridge, UK: Cambridge University Press, 1987), pp. 14–41.

8 See: Philip R. Berke and Timothy Beatley, *Planning for Earthquakes: Risk, Politics and Policy,* op. cit., and David R. Godschalk, *et al., Catastrophic Coastal Storms: Hazard Mitigation and Development Management,* op. cit.

9 See: Robert E. Deyle, "Conflict and the Role of Planning and Analysis in Public Policy Innovation," *Policy Studies Journal* 22, no. 3 (Autumn 1994): 457–473; and Peter J. May, "Policy Learning and Failure," *Journal of Public Policy* 12, no. 4 (1992): 331–354.

10 Two regression equations were employed. For one, the dependent variable was the commitment of elected officials. For the second, the dependent variable was commitment of staff. The units of analysis were local governments in Florida and New South Wales for which cases were weighted to give each set of data equal weight in regression calculations (total n = 125). For the first regression (adjusted R2 = .17, p <.05), factors with noteworthy standardized coefficients are catastrophic losses (B = .33, p <.01), constituency demands for hazard mitigation (B = .27, p <.01), and population growth rate (B = .14, p <.10). For the second regression (adjusted R2 = .23, p <.05), factors with noteworthy standardized coefficients are catastrophic losses (B = .17, p <.10), constituency demands for hazard mitigation (B = .48, p <.01), government expenditures per capita (B = −.16, p <.10). The average effect of being a Florida jurisdiction as compared to being a New South Wales jurisdiction was .15 for each equation. However, this approached statistical significance only for the elected officials' regression (p <.10). Other variables in each regression were additional psychological variables (prior losses, potential losses) and practical factors (population, staff per capita, and staff professionalism). The modest coefficients of variation for these regressions indicate that there is substantial variation in commitment not accounted for by the factors we examined.

11 We found a strong positive effect of technical assistance in leading to increased levels of commitment (Beta coefficients of .23, p <.05 for elected officials, and .23, p <.05 for staff) for regression models that included this variable along with other factors modeled in the preceding analysis.

12 William A. Anderson and Shirley Mattingly, "Future Directions," in *Emergency Management: Principles and Practice for Local Government,* eds. Thomas E. Drabek and Gerard J. Hoetmer (Washington, D.C.: International City Management Association, 1991), pp. 311–336, at 318.

13 Janet A. Weiss and Mary Tschirhart, "Public Information Campaigns as Policy Instruments," *Journal of Policy Analysis and Management* 13, no. 1 (Winter 1994): 82–119, at 83.

14 For discussion of this and related consequences of information campaigns, see Janet A. Weiss and Mary Tschirhart, "Public Information Campaigns and Policy Instruments," op. cit.

15 Information strategies employed in Florida are described in Robert E. Deyle and Richard A. Smith, *Storm Hazard Mitigation and Post-Storm Redevelopment Policies*

(Tallahassee, FL: Department of Urban and Regional Planning, The Florida State University, 1994).

16 Local governments in New South Wales averaged 3.02 (s.d. = 1.69) and local governments in Florida averaged 4.04 (s.d. = 1.80) on our seven-point scale for respondents' ratings of the awareness-building efforts of their jurisdiction. This difference is statistically significant (p <.01) for two-tailed t-test of differences.

17 For discussion of these aspects of planning see: Paul Davidoff, "Advocacy and Pluralism in Planning," *Journal of the American Institute of Planners* 31, no. 4 (1965): 331–338; and John Forester, *Planning in the Face of Power* (Berkeley, CA: University of California Press, 1989).

18 See: Andreas A. Faludi, *A Decision-Centered View of Environmental Planning* (Oxford, UK: Pergamon, 1987); and, Ernest R. Alexander, *Approaches to Planning,* 2nd edition (Philadelphia, PA: Gordon and Breach, 1992).

19 See: Edward J. Kaiser, David R. Godschalk, and F. Stuart Chapin, Jr., *Urban Land Use Planning,* 4th edn (Urbana, IL: University of Illinois Press, 1995).

20 This is the same procedure employed to rate the quality of regional policy statements in New Zealand that formed part of our analysis in Chapter 6. The coding involved looking for the presence or absence of a series of potential hazard management goals and policies. For an explanation of the coding of the Florida data see: Philip R. Berke and Steven P. French, "The Influence of State Planning Mandates on Local Plan Quality," *Journal of Planning Education and Research* 13, no. 4 (1994): 237–250.

11

PROSPECTS FOR COOPERATIVE INTERGOVERNMENTAL POLICIES

Future directions for environmental policy revolve around enduring issues about the role of governments in preventing environmental harms and averting environmental catastrophes. A rethinking of approaches to environmental regulation has been stimulated by acknowledgment of the limitations of current policies for these purposes. The intergovernmental dimensions have taken on cogency with the recognition of the importance for environmental sustainability of the decisions that local governments make about land use and development. Many argue that future directions for environmental policy include less emphasis on regulatory prescription and greater reliance on local governments as partners in pursuing paths to sustainable futures. Bringing this about requires a rethinking of the ways in which higher-level governments influence the decisions of local governments. At issue is how intergovernmental policies can be better designed and implemented to increase shared commitment among different layers of government to environmental goals and to foster practices that promote, rather than undermine, sustainability.

This book has addressed what we label as "cooperative intergovernmental policies" as one promising direction for structuring the intergovernmental approach to environmental management. These policies assume that lower-level governments are willing to serve as regulatory trustees and, with appropriate guidance and assistance from above, are capable of figuring out ways of reaching desired state or national environmental goals. The emphasis is on prescribing goals and planning processes, while supporting efforts of lower-level governments to figure out best practices for reaching those goals. Current ideological preferences for fewer dictates from above, recognition of the conflicts engendered by reliance on intergovernmental coercion, and demands for flexibility by local government officials all point to emphasis in the future on cooperative forms of governance.

Although this approach appears to have much promise, basic questions remain about the ability to bring about the desired intergovernmental cooperation and the efficacy of the approach. In this book, we have addressed these questions by examining leading examples of this approach to environmental management in New Zealand and to hazard management in New South Wales, Australia. In both

213

of these cases, major policy changes were undertaken with aims of fostering co-operative governmental partnerships for environmental or hazards management. A central aspect of each is empowering local governments to devise paths for managing the environment in a sustainable way. Where appropriate, we have drawn a contrast with the more coercive and prescriptive approaches used in corresponding programs in the United States as exemplified by Florida's widely heralded growth management program.

Both cooperative and coercive forms of intergovernmental policies present dilemmas that are discussed in this chapter. Given these dilemmas, the principal challenge in designing intergovernmental policies is to get the policy mix right. This takes on special importance in the environmental arena where the choices by local governments about land use and development management have noteworthy implications for sustainability. These choices can profoundly affect the economic base of communities, the quality of the environment, and the vulnerability of communities to natural hazards.

This chapter summarizes key findings and draws implications from them for the design and implementation of intergovernmental policies. The chapter begins with an assessment of the policy innovations examined in the preceding chapters and summarizes the lessons we draw about cooperative intergovern-mental policies. This provides a foundation for discussion of their long-term potential and the general applicability of key features of these policies. Recognition of the limitations of cooperative intergovernmental policies leads to consideration of potential improvements in policy design and implementation.

ASSESSING THE POLICIES

The environmental and hazard-management experiences discussed in this book are best considered a set of policy experiments involving contrasting inter-governmental regimes. The lessons we draw from the experiences are necessarily tentative because of gaps in data collection, difficulties of making cross-national comparisons, and the fact that the New Zealand regime was only in its third year of implementation at the time of our fieldwork. Nonetheless, these experiments have noteworthy implications for rethinking the intergovernmental dimensions of environmental management.

Policy evolution, implementation, and learning

The rich policy histories discussed in Part I of the book document the struggles that policy-makers faced in establishing and refining policies. State policy-makers in Florida sought over time to devise a politically acceptable means for instituting a strong growth management program that would not engender substantial backlash by local government officials. The reforms that were under-taken in New Zealand in the late 1980s and early 1990s are dramatic attempts to reallocate responsibilities for the management of natural and physical resources among different layers of government. These reforms have attracted international attention because of the uniqueness and scale of the undertakings. The merits approach to state flood policy in New South Wales was crafted in 1984 to overcome the backlash that the prior prescriptive policy created. In the subsequent decade, state administrators made adjustments in policy guidance in order to accommodate the diversity of local government situations.

Relevant central or state government agencies in each setting have adopted approaches to dealing with lower-level governments that are generally consistent with the coercive or cooperative character of the respective intergovernmental policies. The approach of the Department of Community Affairs in Florida toward local governments stands out as being formal and legalistic. This is a stark contrast to the more informal and facilitative dealings with local government officials by the Ministry for the Environment and the Department of Conservation in New Zealand, and by the Department of Public Works and the Water Resources Department in New South Wales.

However, inconsistencies are evident in the translation of broad policy objectives of cooperative regimes into practice. Some of the state agencies in New South Wales and central government agencies in New Zealand are not expending much effort in implementing the relevant policy. In addition, there is variation in the degree to which these government agencies, especially among regional offices, have incorporated a facilitative style into their day-to-day dealings with local governments. Although problems of obtaining consistent implementation are common in the intergovernmental implementation of policy, the inconsistencies are potentially more problematic for cooperative intergovernmental regimes because they contribute to confusion and mistrust on the part of local governments. We attribute the inconsistencies to a mix of internal and external forces for which the commitment and capacity of the state and national agencies are especially important factors in shaping the way that they approach local governments.

As policy-makers in each setting have grappled with the various challenges of policy design and implementation, different types of policy learning are evident. State policy-makers in Florida showed evidence of political learning as they fine-tuned the growth management program in response to complaints from local government officials. Regional agencies were given less power to prescribe local policy choices, and new provisions were added to increase local government protection of property rights. Under the merits flood policy, state agency personnel in New South Wales have been at the forefront of innovative ways of managing flood hazards and in working with local governments to alter the mix of regulatory measures. This reflects learning by state and local officials about the adaptation of flood-reduction techniques to different local circumstances. Government officials at all levels in New Zealand are on a steep learning curve in their efforts to fully realize the promise of the extensive reforms in local government structure and resource management.

Despite such learning by policy-makers in each setting, noteworthy obstacles remain in the way of accomplishing policy objectives. Administrators of Florida's growth management program face the challenge of maintaining momentum in the face of continuing criticisms of state actions and fiscal uncertainty. Substantial resources are required for the program at state and local levels, presenting challenges for keeping it in place. The situation facing administrators of flood policy in New South Wales is in many ways the inverse of the challenges posed to Florida's growth management program. The flood policy appears to be well accepted by local government officials, but that is partly because state officials have not been assertive in promoting the program within reluctant flood-prone localities. In addition, the lack of integration of the flood program with other environmental policies is a stark contrast with comprehensive planning undertaken by local governments in Florida. The principal challenges in New Zealand are making the policy work with limited local resources and limited central government involvement. The early evidence is in many ways promising. Regional governments are playing important facilitating roles, albeit with notable exceptions in some areas. In addition, regional and local government personnel appear to be genuinely grappling with the new modes of environmental management that are called for under the new regime.

Strengths and limitations

This book sheds light on the strengths and limitations of cooperative and coercive intergovernmental regimes as they relate to local planning processes, compliance, and innovation. The coercive intergovernmental approach, as implemented in Florida by state agencies with some degree of flexibility, has advantages in achieving compliance by local governments with procedural provisions. Facing the threat of sanctions, local governments produced plans in a timely manner. With the help of additional funds and after negotiating agreements specifying conditions for compliance, virtually all of the local government plans were brought into compliance with the state's planning requirements. Local governments in New South Wales have had lower rates of compliance in preparing floodplain management plans, despite strong incentives to adhere to the state's recommended process. It is too early to firmly gauge how compliance by local governments in New Zealand will play out, but the evidence to date suggests a future pattern that looks more like that found in New South Wales than in Florida.

With a coercive intergovernmental mandate, state (or national) governments are better able to get local governments marching to the state (or national) government's tune. However, there is a potential that a coercive mandate will be viewed by local government officials as too heavy-handed. Such perceptions can, in turn, lead to a backlash that will unravel the whole policy. This happened in New South Wales under the state's prescriptive mandate that prohibited local government approval of development within floodplains. That policy unraveled when several local governments refused to participate and others took actions that undermined the policy. State policy-makers in Florida have had to respond to the frustration of local government officials to what they perceive as heavy-handed state agency actions by supplying additional funds and seeking negotiated settlement of differences.

Our analyses show that procedural prescriptions about local planning alter the character of planning processes. There was greater attention to environmental considerations in both New Zealand and New South Wales after the introduction of policies with new planning provisions. These effects, however, have more to do with procedural prescription about the planning process than the type of policy mandate. With respect to the latter, local governments undertaking planning under the cooperative intergovernmental approach in New South Wales had higher-quality floodplain management plans than the corresponding hazard-management components of local plans produced by local governments in Florida. We think these differences in plan quality result from a combination of greater time to prepare plans, more extensive resources for each plan preparation, and the focused nature of floodplain management plans.

One of the key differences in the policies we study is with respect to fostering innovation in new approaches to planning, new forms of regulatory instruments, or the re-combination of regulatory provisions to better match local circumstances. While such innovation in itself need not result in better environmental management, recognition of the limitations of past approaches has clearly renewed interest in seeking new or more appropriate regulatory solutions. The resource management and associated government reforms in New Zealand sought such innovation with a particular emphasis by central government officials in promoting local use of new forms of regulatory instruments. The flood policy in New South Wales sought local determination of appropriate flood standards while seeking innovative approaches to establishing standards. Florida's growth management program was different in that it imposed state-formulated innovation (e.g., in consistency and concurrency requirements) on local governments, and sought compliance by local governments in implementing the state-devised innovations.

Given the approaches to encouraging local innovation, it is not surprising that we find differences in the nature of innovation among the three policies. In general, there was more evidence of innovation in local planning and plan content in New Zealand than elsewhere. However, New Zealand's central government officials have indicated disappointment with the pace of innovation and local planners have expressed desires for more prescription about acceptable plan and regulatory provisions. One clear point is that, if it is to occur at all, it takes time for innovation and an active effort to promote new ideas. In New South Wales, many local governments fell back on traditional flood standards rather than developing innovative ones. The problem seemed to be that local officials were reluctant to adopt standards that had not passed prior legal challenges.

In both New Zealand and New South Wales there was a much stronger sense than in Florida among local governments of an ability to devise new approaches to environmental or hazards management. Local planners in Florida felt inhibited by the prescriptive and coercive provisions of the state's growth management program. They felt much more restricted in their ability to innovate. From these experiences, we conclude that innovation can be facilitated by cooperative intergovernmental regimes but it will not occur without substantial follow-through with technical assistance and attention to concerns by local government officials about the legal standing of innovative approaches.

Dilemmas of cooperative and coercive regimes

Both coercive and cooperative forms of intergovernmental policies present dilemmas. The dilemma for coercive regimes is that in bringing about procedural compliance by lower-level governments and forcing them to pursue state-initiated innovations, they may straitjacket local governments that want to develop their own innovative solutions to environmental problems. By emphasizing procedural over substantive compliance, coercive intergovernmental regimes can foster a cookie-cutter compliance mentality (i.e., token, formula-based compliance) in meeting state-defined deadlines and objectives. Nonetheless, procedural compliance by local governments with planning processes and plan requirements is important. Such processes provide at least a minimal set of local government provisions for achieving environmental regulatory objectives, and the completion of a plan provides a means to involve local constituencies who in turn can create political support for more extensive policies. One key caveat for coercive intergovernmental regimes is that coercive regimes run the risk of being undermined by a substantial backlash by local government officials.

The dilemma for cooperative regimes is that they foster local ownership of environmental management programs, but they suffer from gaps in procedural compliance because of reluctance by some local governments to follow the prescribed policy process. Local ownership is important because it provides the constituency base for fashioning acceptable land-use and development rules, thereby enhancing prospects for implementation and effective environmental management. The gaps are troublesome, however, because they clearly create spotty compliance with state (or national) desires and because such gaps are difficult to address. Cooperative regimes lack the type of sanctions found in coercive policies to motivate recalcitrant local governments and instead must rely on incentives that are imperfect for this purpose. One key caveat for cooperative intergovernmental regimes, discussed in what follows, is that the flexibility provided by cooperative regimes opens the door to potential capture of local decisions by development interests.

Capture potential and sustainability

This potential for capture raises two challenges that have been posed about decisions local government officials make under cooperative intergovernmental regimes. One is that they cannot be trusted to make socially beneficial decisions about land use and development in environmentally sensitive areas. Those that argue this perspective see elected officials in local government as either single-mindedly pursuing economic development or dominated by parochial development interests. A second, related challenge is that there is an uneven playing field in making development decisions so that they do not represent relevant stakeholders in an equitable way. The solution to either of these challenges is for higher-level governments to step in and exert strong controls over decisions by local governments, thereby reverting to coercive intergovernmental policies.

Respondents from local governments to our surveys in New South Wales and in New Zealand report a change under cooperative regimes in the approaches that they take to managing natural hazards. In particular, the survey data for local governments show greater flexibility in rules and greater willingness to negotiate development decisions. This is consistent with the notion of case-by-case decisions on the merits of development that is an explicit goal of the New South Wales policy, and is consistent with the emphasis in New Zealand on local government consideration of environmental effects as the basis for sustainable management of natural and physical resources. Our case studies in New South Wales show that the locally derived regulations under the merits flood policy provide flexibility because they relate better to local situations. Negotiation typically takes place over the means for adhering to the rules, revolving around best practices to meet the locally specified standards, rather than over the standard per se. The positive aspect of this flexibility is that it resulted in local regulations that were more acceptable than in the past to various community constituencies, although this was not uniformly the case.

As important for future sustainability is that local governments in New South Wales and in New Zealand altered their approaches to managing natural hazards under the cooperative regimes. They were more likely to rely on building and land-use rules than in the past, resulting in policies that generally lessen potential for long-run, catastrophic losses. The shift in the regulatory mix is explained by the empowerment of professionals who advocated such solutions under cooperative planning processes, and by the retrenchment in state or national funding for new structural protection schemes such as levees. The latter was particularly important in New Zealand.

Yet, the flexibility provided under the cooperative policies has the potential to result in different directions. This was evident for the case studies in New South Wales and in New Zealand that contrasted leading and lagging approaches to sustainable management of flood hazards. The forces that railed against strict regulations in New Zealand and New South Wales are still evident under their flexible systems. Just as some communities dutifully applied strict regulations in the past, so too under cooperative regimes some communities appear to have developed comprehensive programs to reduce flood hazards. Yet, others appear to be doing little in the age-old belief that growth, even in hazardous areas, is good for their community.

In sum, there is a range of possible outcomes under cooperative intergovernmental policies for influencing local management of risks posed by natural hazards. These relate both to the responses of different localities in managing the risks and to the outcomes of those efforts in reducing exposure. The positive points are that cooperative policies seem to allow flexibility in local decision-making to permit choices that lessen risks of long-run catastrophic losses and improve resilience against more frequent events. We consider these to be important aspects of sustainability with respect to natural hazards. Yet, the cooperative policies also leave the door open for non-compliance and for choices that do not promote long-run sustainability. Our statistical analyses show that a variety of factors contribute to these differential outcomes. The commitment of elected officials and staff to these aspects of environmental management loom large in the overall equation.

PROSPECTS FOR COOPERATIVE POLICIES

Case histories of public policies are replete with descriptions of policies that were short-term successes but longer-term failures, and others that appeared to be failures in their early stages but became more successful as they matured. Sometimes policy success or failure can be closely tied to a set of situational factors relating either to the substance of the policy, or to the setting in which the policy is being carried out. In this section, we address different aspects of the future prospects of cooperative intergovernmental policies.

One can think about the future prospects for policies over the course of decades, rather than years, in a variety of ways. One way is to consider the viability of a policy as an instrument of government given potential challenges to the policy and changing circumstances. This is difficult to address because much depends on the nature of the challenges and circumstances as well as on the skill of supporters of the policy in responding to the challenges. One of the remarkable aspects of the Florida program is that it has survived strong legislative and legal challenges while maintaining its basic features. The merits flood policy of New South Wales has been free of serious challenges and has been well accepted by local governments in the decade since its inception.

The occurrence of major natural disasters is an especially relevant circumstance for the policies we consider. Our analysis shows that catastrophic losses are an important factor in mobilizing constituency demands for dealing with risks posed by natural hazards. This, in turn, affects commitment of local government officials to address the problem. The more subtle point about such occurrences is that they can be expected to be more beneficial for advancing hazard management under cooperative regimes than under coercive regimes. This is because the catastrophic event can serve as a stimulus to the local governments that previously chose not to address the risk posed by a potentially large, low-probability event. The lesser impact under coercive intergovernmental regimes stems from the fact that most (if not all) local governments will have already taken some precautionary action, if for no other reason than to avoid the sanctions associated with the coercive policy. But, as discussed in the previous chapter, simply waiting for catastrophic events is not good policy. In the intervening period potentially unsafe development is allowed to continue in hazard-prone areas, thereby increasing risks and making it more difficult to address the problem.

It is clear that either coercive or cooperative regimes can unravel over time. Commitment may erode under coercive intergovernmental mandates, particularly as monitoring and enforcement ease. Complacency is a key challenge to coercive mandates as they mature. National or state agencies charged with monitoring and enforcement may take for granted compliance by local governments. If changing local circumstances alter the calculations about compliance by local governments in favor of non-compliance, state-level complacency could be a problem. Similarly, cooperative intergovernmental regimes can unravel if the interests of the various levels of government diverge over time. National or state authorities may find themselves faced with increasing gaps in participation by local governments, thereby leading toward coercion in an effort to spur action by the recalcitrant governments. The point is that either form of policy mandate requires adjustments over time, particularly in responding to changes in the willingness of local governments to comply with the dictates or wishes of higher-level governments.

A second way of thinking about policies over the long term is to consider how a given set of policy impacts will evolve over a decade or more, assuming that the policy and external circumstances remain fairly constant. We obtain glimpses of this from the contrasting experiences of nearly a decade for the coercive intergovernmental policy in Florida with those of the cooperative policy in New South Wales. The coercive policy seemed to have a greater impact in inducing compliance by local governments and in stimulating constituency demands for local governments to address risks posed by natural hazards. But, we also observed that the commitment of elected officials and staff of local governments (particularly the latter) increased under cooperative regimes. Our data did not include direct measurement of change in commitment under Florida's coercive regime. As a consequence, we are unable to evaluate the assertion that cooperative intergovernmental policies have greater long-run promise because they foster greater local ownership of programs, and thereby build commitment to environmental management.

A third way to think about policies over longer periods of time is thinking about policies as endogenous processes that foster new circumstances. These, in turn, create new demands for policy change. This is consistent with notions offered in the political science literature that policies beget politics.[1] One fairly immediate potential set of impacts is a political reaction to the policy, leading to calls for reform or abolishment. Leaving this aside, three types of impacts seem relevant to this way of thinking about policies over time: mobilization of constituency demands that results from participatory-planning processes;

empowerment of planners and other professionals; and, legitimization of regional governments as important entities in the intergovernmental system.

As discussed in the preceding chapter, regardless of the type of intergovernmental regime, participatory planning processes can have positive effects in mobilizing the demands of key community groups for addressing particular problems. Such mobilization is important because constituency demands are critical factors in influencing the commitment of local elected officials to higher-level policy goals. Yet, how this plays out over time depends on the constituencies that are mobilized for which the potential for capture of local decision-making by development interests cannot be ignored. In Florida and New Zealand, the participatory planning processes have fostered involvement of a number of groups and resulted for some localities in strong grass-roots demands to address environmental problems. This is helpful in focusing attention by elected officials and moving the relevant problem onto the local agenda, but it can also lead to conflict within communities about appropriate solutions. In New South Wales, committees that were established within local governments for the purpose of guiding local floodplain management seemed to have been effective in reaching a consensus among stakeholders about an appropriate course of action. However, lack of the integration of floodplain management planning with broader environmental issues limited creation of community-based constituencies for broader environmental management.

A second impact that can take on a life of its own is the empowerment of planners or other professionals. Planning provisions of either form of intergovernmental policy provide empowerment of particular professions, with planners being empowered under the policies in Florida and in New Zealand and engineers being empowered under the flood policy in New South Wales. Because different professions have distinctive ways of solving problems, one potential set of longer-run outcomes is a shifting of the grounds of debate about local development and sustainability. Where planners become influential, greater emphasis can be expected on land use and development-management solutions to environmental problems. Engineers are more likely to suggest solutions involving structural works or other means of controlling hazards, although in New South Wales there was a strong emphasis on non-structural solutions to flood problems. One reason for this was the extensive role played by a forum of local government professionals and elected officials, formed in the late 1950s for the purpose of lobbying for state and Commonwealth funding for flood programs. Over the years, this has been an important forum for learning about innovative ways of managing flood risks.

A final set of potentially interacting consequences of intergovernmental regimes are the consequences of the legitimization of regional tiers of government. Here we see variation in provisions for the policies we studied. Policy-makers in Florida and New Zealand incorporated strong roles for regional entities, while state policy-makers in New South Wales employed a more *ad hoc* system as needs arose. In Florida, regional planning councils were placed in a hierarchical position *vis-à-vis* local governments and were given few financial or political resources. As a consequence, they retained a precarious situation and have not been very effective. In New Zealand, the devolution of responsibilities for environmental management breathed new life into a newly reconstituted regional tier of government. The legacy of uneven relationships of the predecessor regional entities with local governments could lead to disfranchisement of the newer regional governments, but the evidence to date suggests this is not the case. They are performing useful functions, albeit with notable exceptions for some regions. If regional governments gain stature through this process over time, central government policy-makers may face demands from regional governments for more control over the environmental policies of local governments.

PRECONDITIONS FOR COOPERATIVE POLICIES

Drawing on our understanding of the different policy regimes, we can suggest some of the preconditions for workable cooperative intergovernmental regimes. These are characterized as a set of primary conditions that must be present and a set of facilitating conditions that ease policy enactment and implementation.

Primary conditions

By definition, each form of intergovernmental mandate requires a multi-tiered governmental structure. The relevant tiers will depend on the type of policy and respective responsibilities of different levels of government. Such policies are not limited to federalist systems of governance, as there are many conceivable ways of sorting out responsibilities among different layers of government. We have considered state influence over local governments under federalist systems in Australia and the United States, and national influence over regional and local governments under New Zealand's unitary system of governance.

A second primary condition is a shared set of policy objectives among higher- and lower-level governments. Policy-makers must have sufficient desire to take on environmental problems and to structure an intergovernmental policy around cooperative principles. Where commitment to policy objectives is strong, officials at any level of government have motivation to act in accord with each other. When there is fundamental disagreement over policy objectives or the allowable range of means for meeting them, the cooperative nature of the intergovernmental partnership will be doomed from the start. It is hard to imagine strong divergence over the goal of addressing risks posed by natural hazards. Different levels of government have incentives to protect lives and avert damage if for no other reason than to reduce the costs of disaster relief. However, it is easy to imagine a workable partnership going astray when governmental partners differ over the extent of desired risk reduction, the means for achieving it, or who bears the cost of carrying out relevant programs and disaster relief.

223

The concept of sustainability has entered the policy debate as an objective for environmental management. One of the appeals of the concept, when employed at a fairly abstract level, is that it is sufficiently vague so as to provide an objective for which consensus can be reached. Conflicts arise, however, when efforts are made to be more precise in defining the concept since there is little agreement about various economic, environmental, or social dimensions that the concept potentially invokes. One of the astute decisions by architects of the New Zealand reform in environmental management was to avoid putting a fine point on the vision of sustainability that is the centerpiece of the reform. The legislation establishes sustainable management of natural and physical resources as a vision and specifies the importance of what has become known as an "environmental bottom line" for working toward that vision. But, the legislation leaves it mainly up to officials of regional and local governments to further refine what this constitutes for their jurisdiction and how to achieve it. This makes it possible to endorse the notion of sustainable management without engendering broad conflict over the means of pursuing this goal.

A less evident issue is the extent to which devolution of governmental power to lower levels of government is necessary for cooperative intergovernmental regimes and a hindrance for coercive regimes. Our policy exemplars provide a contrast between a highly centralized system in Florida with much authority resting in state government and a decentralized system in New Zealand with power devolving to lower levels of government. Given New Zealand's history of strong central and local government with weak and fragmented regional government, the regional governments were in a potentially precarious position. The conscious devolution of decision-making to regional (and local) councils provided them with a stronger footing than would otherwise be the case. Indeed, some might argue that a regional government role is essential for filling the partnership void left by the retrenchment of central government involvement in environmental management.

In contrast, coercive intergovernmental regimes require a higher-level authority to carry out monitoring of compliance by lower-level governments and the power of the state to carry out enforcement when compliance is lacking. Because of this, coercive regimes have powerful centralizing influences that reduce the need for a regional role. This is illustrated by the limited role that regional entities have performed in Florida. Despite being assigned important functions, regional planning councils have been weak entities and subject to political attack because of their seeming ineffectiveness. In centralized systems, regional entities need a clear purpose and stable bases of political and financial support in order to thrive. This is illustrated by relatively powerful water resource management districts in Florida and special-purpose regional entities in New South Wales. These have strong roles in floodplain management that provide relatively clear benefits, and neither is dependent on the state government for funding or political support.

The preceding discussion suggests that the extent to which regional government roles can be important for functioning of either cooperative or coercive intergovernmental regimes depends on two factors. One is the extent of the devolution of power from higher-level governments to lower levels. As the New Zealand case illustrates, regional governments can play important facilitating roles under cooperative regimes under devolved systems. A second factor is the extent to which regional entities derive separate bases of financial and political support. While this evidence supports the case for a regional role, we can imagine other workable forms of coordination and assistance not involving the regional tier of government.[2] One example is regional trust or compact among local governments that calls for coordination of policies, shared data about environmental effects, and perhaps some form of joint funding.

Facilitating conditions

One point to keep in mind is the variation in the specifics of cooperative intergovernmental regimes. Cooperative environmental policies that ask local governments to manage on the basis of environmental outcomes, rather than on the basis of particular uses, place noteworthy demands on all levels of government. Such "effects-based" management is a cornerstone of the New Zealand approach and an important component of the New South Wales merits flood policy. Successful implementation of this approach necessitates a means for assessing environmental effects and a capability to devise solutions that have discernible impacts in reducing environmental harms or in protecting valued resources. These place a premium on creation of databases and provision of technical assistance.

Cooperative intergovernmental regimes are more suitable to situations where there is normative commitment by local government officials to higher-level policy goals. In the absence of this, strong incentives supplied by higher-level governments are essential for inducing participation and building the requisite normative commitment. Examples of incentives include funding for participation of local governments and removal of the legal uncertainty that local officials anticipate if they devise new ways of managing the environment. However, these may be insufficient for substantially closing gaps in participation, as illustrated by the experience in New South Wales where both funding and legal incentives were strong.

The approach of national or state agencies to local governments is also important. Under a cooperative intergovernmental mandate, the challenge for higher-level government agencies is to provide leadership in facilitative and non-interventionist ways. The amount and nature of assistance provided by higher-level governments is of critical importance to achieving successful outcomes for cooperative mandates. In New Zealand, the greatly reduced role of central government has resulted in confusion among local officials about how to achieve the policy intent and delays in policy implementation. These criticisms did not arise in New South Wales where staff of relevant state agencies took a more hands-on, consultative approach in dealing with local governments. Our statistical findings show that the commitment *and* capacity of relevant

225

implementing agencies need to be strong in order to foster the type of informal, facilitative approach to local governments that is required of cooperative regimes. This puts a premium on having well-funded and committed higher-level government agencies, in addition to having requisite resources for assisting lower-level governments.

Cooperative regimes also require greater cross-policy consistency than do coercive regimes. Inconsistencies arise when local government officials face both a cooperative policy that promotes local innovation and a coercive mandate that prescribes particular actions for a given aspect of environmental management. Unless there is clear evidence that it is not being enforced, compliance will tend to be driven by the latter since the prescription is clear and sanctions can be imposed. We noted that inconsistencies in New Zealand between the prescriptive flood provisions under the *Building Act* and the less-prescriptive corresponding features of the *Resource Management Act* impeded local government innovation in dealing with aspects of flooding. However, in New South Wales inconsistencies between prescriptive state planning directives and the merits flood policy were not problematic because the former were not well known (even among relevant state personnel) and were not enforced.

Related to this is the role of legal culture and supporting legal structure in shaping policy implementation. Coercive policies engender legal formalism by virtue of strict review processes and the stakes involved in applying sanctions. Cooperative intergovernmental policies are hindered by legal formalism because it undermines the spirit of the intergovernmental partnership. In addition to expected differences in legal formality in the settings we studied, there were also noteworthy counter-currents. State administrators in Florida sought to reduce legal formalism by negotiating agreements with local government officials over the specifics of policy compliance. However, the agreements took on a legalistic form given the fact that failure to meet them could result in sanctions. Policy-makers in New Zealand sought to encourage local innovation in sustainable management, but legal formalism acted as a restraint during a central government agency review of policy statements produced by regional governments. In addition, the reliance of local governments on the decisions of the national Planning Tribunal to provide guidance about acceptable planning practice fostered legal formalism in the planning process. The lesson here is that breaking away from legal formalism is difficult where there are strong legal institutions and traditions.

Part of the participatory planning processes established under intergovernmental regimes is active involvement of a range of community groups. In particular, we have shown that the mobilization of community groups is an important aspect of building the commitment of local officials to take action in addressing hazard-prone areas. This means that the existence and viability of local interest groups can facilitate successful policy outcomes, but at times conflict may increase or special interests may capture decision-making. Risks such as those posed by natural hazards that are broadly distributed and perceived as being remote often do not generate broad-based community involvement until a catastrophic event occurs. Because of this, well-crafted, targeted information campaigns and active efforts to mobilize relevant community and professional groups are important. Clearly, the nature of this mobilization can make a difference both in terms of the perceived fairness of decision-making and the eventual outcome of decisions.

The tractability of problems, defined with respect to the ease of addressing the problem, is a commonly cited factor in shaping the success of implementation.[3] Our findings suggest that problem tractability is more important in easing implementation under cooperative regimes than under coercive regimes. As problems become more difficult to address, it is almost tautological that policy-makers will be more resistant to address them without a strong push to do so. Among the settings we studied, Florida presented the least tractable set of problems as a result of the extreme hurricane risk and hazard-prone areas already being highly developed. Thus, it is not surprising to find limited local government action in addressing these problems under the non-coercive intergovernmental regime that existed from 1965 to 1984. In contrast, there was extensive local government participation under the coercive intergovernmental regime that was instituted in 1985.

APPLICABILITY OF THESE LESSONS

Although the topical focus of this book is the management of natural hazards, the selection of policy experiences and findings clearly involve broader issues concerning environmental, growth, and natural resource management. The lessons provided in this chapter would be similar had we considered more fully these broader topics. Because provisions relating to natural hazards were not the major foci of the policies in Florida or in New Zealand, we found ourselves often talking with interviewees more generally about growth management (Florida) or environmental management (New Zealand). In these conversations and related survey findings for local governments, there was little distinction between issues that arose for managing the broader policies and those for hazards policies.

More generally, each of the policies in this book can be considered as a form of development policy that involves decisions by local governments about future economic growth. Economic development decisions about such items as infrastructure, housing, and parks entail similar debates and tradeoffs as decisions about land use and physical development. These can have profound effects on economic vitality, environmental quality, and vulnerability to natural disasters. The main difference is that local government officials are likely to be more supportive of economic development than they are of environmental regulation. The compelling logic of cooperative intergovernmental regimes that have been the foci of this book is reinforced by the argument that development policies work better when local governments are the central players and higher-level governments are supporting entities.[4]

The ease with which features of cooperative intergovernmental regimes can be transferred to policies in other settings depends on the fulfillment of the conditions discussed earlier in this chapter. Both possibilities and challenges are evident when considering the transferability of such concepts to the United States. The closest parallel in governmental functions, as illustrated by the selection of Florida in the United States for our comparison, is state efforts to influence the decisions that local governments make about land use and development. Nearly a dozen states have put in place programs like that of Florida directing local governments to manage growth. But, the unwillingness of state officials to take on environmental problems and establish cooperative policies could pose a major obstacle. The fact that not more states have addressed growth management reflects the hostility that many local officials have toward state-level dictates, a backlash about the type of heavy-handed approach taken in Florida, and strong grass-roots concerns over governmental abuse of property rights. One of the potentially appealing aspects of cooperative intergovernmental regimes is that they offer a means for diffusing these concerns by establishing partnerships that allow flexibility in developing locally derived approaches to environmental and growth management.

One major constraint in transferring key concepts from the policy framework in New Zealand to American states is the absence of sub-state regions that correspond to natural boundaries. County-level governments do not fulfill this purpose since they do not follow natural boundaries. Other forms of special-purpose districts, such as water resource districts, are too narrow in function and do not exist in many geographic areas. Creation of a new regional tier of government with natural boundaries, as was essentially done in New Zealand, would prove problematic for the reasons that we cited in Chapter 6 about the precariousness of regional entities.

Another constraint is the presence of an adversarial legal culture. The American system is exceptional for its procedural and legal complexity with respect to regulatory programs, and these apply as well to the intergovernmental policies. We have shown that legal formalism can undermine the facilitative approach to dealing with local governments that is essential for conveying the spirit of cooperative regimes. Breaking through these complexities and associated adversarial climate is likely to be a noteworthy challenge. Our analysis of inter-governmental implementation in Chapter 5 suggests it is possible to overcome these obstacles. However, doing so requires high degrees of commitment and strong capabilities on the part of those agencies assigned responsibility for over-seeing implementation of cooperative policies.

GETTING THE POLICY MIX RIGHT

Neither coercive nor cooperative intergovernmental policies are ideal. Each presents dilemmas discussed earlier in this chapter for which a fundamental constraint is the variability in commitment of local government officials.

Cooperative regimes fall short as less committed governments make half-hearted efforts to carry out higher-level wishes or ignore their requests. Coercive regimes can backfire if they are perceived as being too heavy-handed. In addition, more committed local governments feel constrained by the restrictions of prescriptive and coercive mandates. Given the variability in situations, it is clear that some mix of provisions is necessary. The challenge, of course, is getting this mix right. We conclude this book with a consideration of different aspects of this mix with particular attention to the redesign of cooperative intergovernmental policies.

There are several levels for considering modification of intergovernmental policies. The most basic is to consider changes in the specified governmental roles by respecifying the responsibilities of governmental partners. In order to overcome the gaps created by uncommitted local governments, it might be more effective to have states take direct responsibility for a minimum base level of regulation. Incentives could then be provided for local governments to develop additional regulatory programs with provisions that are consistent with shared goals for environmental management. For example, a state government might issue permits for construction in hazard-prone areas and local governments be granted incentives to create land-use and regulatory controls that are better matched to their situations. The local rules would supplement but not supplant the state permit process. Under this scheme, local governments that have the will and capacity to initiate environmental management programs that are more comprehensive will not be constrained. The critical issue for this scheme is the willingness of local governments to accept a base level of state regulation in return for the option to be allowed to undertake more extensive locally developed programs.

A different approach to the redesign of cooperative intergovernmental frameworks is to create extra-governmental mechanisms that compel, short of coercion, participation of local governments in state (or national) programs. We have shown that one key to gaining the commitment of local officials is mobilizing relevant constituencies that in turn create pressures to take action. Cooperative policies might include features that empower third parties to become advocates aimed at mobilizing broader constituencies. For example, funding might be provided as part of a given state's cooperative regime to create a consortium of groups to become "sustainability watchdogs."[5] This works to build normative commitment to policy objectives while recognizing that co-operative policies entail socialization processes that extend beyond governmental entities. This approach also incorporates key tenets of the informational and participatory-planning processes discussed in the previous chapter as being critical to overcoming the commitment conundrum. Much, of course, depends on the effectiveness of the strategies employed by the third parties, the credibility of these organizations in making claims about governmental performance, and the make-up of the groups that perform these functions.

Short of fundamental change in the design of cooperative intergovernmental regimes, key features of such policies can be strengthened. The obvious points are to emphasize the type of incentives (e.g., funding, legal indemnity provisions), informational programs, and planning processes that work to enhance commitment of local governments to policy objectives. In addition, efforts can be made to enhance the commitment and capacity of relevant state agencies in order to foster facilitative dealings with local governments. And, greater consistency in policy intent can be sought by reducing conflicts with related policies that are prescriptive and employ legal formalism. All of these are important aspects of the overall performance of cooperative regimes that have been discussed in this book.

Deadlines for policy participation of local governments in producing plans or meeting other procedural requirements also need to be re-thought. These have little meaning for cooperative policies because of the absence of strong sanctions for failing to meet the deadlines. Instituting hard-and-fast deadlines and strong sanctions for failing to meet them can be counter-productive, since these turn cooperative policies into coercive ones. Imposing firm deadlines without meaningful consequences does little good. The key is establishing deadlines that take into account variation in the commitment and capacity of local governments. These are best established as negotiated deadlines that provide reasonable opportunities for local governments to come to terms with policy objectives and provide time for state officials to provide the necessary technical assistance. Failure to meet the negotiated deadlines then becomes grounds for imposition of effective sanctions. This builds on the strengths of cooperative regimes by allowing for efforts to build normative commitment and by giving local governments the opportunity to develop more than cookie-cutter policies based on prescribed templates. Yet, this scheme also incorporates necessary elements of coercion when procedural compliance is lacking by motivating action with meaningful sanctions.

A rethinking of implementation strategies is also relevant. Greater targeting of outreach efforts to those jurisdictions that are incapable or unwilling to participate would seem to be required. Mandating government agencies often have sufficient information to predict these situations at the time of initial policy implementation, which means that plans for targeting can be devised early on. Different approaches can be employed by dealing leniently with local governments that make good-faith efforts to comply with procedural prescriptions, while putting pressure on those that are reluctant, and providing technical assistance to those that are incapable.[6] This type of enforcement strategy makes most sense if the deadlines for meeting procedural prescriptions are negotiated.

This discussion suggests that there are a number of improvements that can be made to the design and implementation of cooperative intergovernmental policies. The solutions are not simple, nor are they straight-forward. Intergovernmental cooperation cannot be simply legislated, for much depends on the mutual trust of the respective partners and shared willingness to tackle common problems like the aspects of environmental management discussed in this book. Finding the right mix may prove to be elusive, but part of the search is a willingness to look for lessons outside one's national boundaries. We hope that our efforts in this book will stimulate further attention to the intergovernmental dimensions of environmental management and foster additional efforts to draw cross-national lessons.

NOTES

1 The seminal works in pointing this out are Aaron Wildavsky, "Analysis of Issue Contexts in the Study of Decisionmaking," *Journal of Politics* 24, no. 4 (November 1962): 717–732, and Theodore Lowi, "American Business and Public Policy: Case Studies and Political Theory," *World Politics* 16, no. 4 (July 1964): 677–715.

2 For discussion of different mechanisms to achieve coordination under decentralized systems see: James L. Sunquist with the collaboration of David W. Davis, *Making Federalism Work, A Study of Program Coordination at the Community Level* (Washington, D.C.: Brookings Institution, 1969); and Donald Chisholm, *Coordination Without Hierarchy, Informal Structures for Multiorganizational Systems* (Berkeley: University of California Press, 1989).

3 See: Daniel A. Mazmanian and Paul A. Sabatier, *Implementation and Public Policy* (Glenview, IL: Scott, Foresman and Company, 1983), pp. 21–25.

4 For this argument, see the discussion of development policies and governmental roles in Paul E. Peterson, Barry G. Rabe, and Kenneth K. Wong, *When Federalism Works* (Washington, D.C.: The Brookings Institution, 1986), pp. 12–20.

5 For related discussion of the prospective roles of non-governmental organizations in environmental regulation see: Ian Ayres and John Braithwaite, *Responsive Regulation, Transcending the Deregulation Debate* (New York: Oxford University Press, 1992) and DeWitt, John, *Civic Environmentalism: Alternatives to Regulation in States and Communities* (Washington, D.C.: Congressional Quarterly Press, 1994).

6 This is similar to flexible regulatory enforcement approaches discussed for private-sector regulation. See Robert A. Kagan, "Regulatory Enforcement," in *Handbook of Regulation and Administrative Law*, ed. David H. Rosenbloom and Richard D. Schwartz (New York: Marcel Dekker, Inc., 1994), pp. 383–422; and John T. Scholz, "Managing Regulatory Enforcement in the United States," in *Handbook of Regulation and Administrative Law*, ed. David H. Rosenbloom and Richard D. Schwartz (New York: Marcel Dekker, Inc., 1994), pp. 423–463.

APPENDIX
Methodological notes

One of the key goals of this book is to provide an empirical basis for examining experiences under different intergovernmental regimes for environmental management. For the selected settings in Australia, New Zealand, and the United States, this entailed collection of data involving various levels of government, case studies of experiences by local governments in hazard management, interviews with officials of state government agencies (in New Zealand, central government agencies), and construction of a set of measures of key concepts employed in the research. This appendix summarizes the approaches to conducting surveys of local governments, carrying out case studies, interviewing government personnel, and constructing measures of key concepts. The comparative research design and rationale for selecting the policies and settings we study are discussed in Chapter 1.

SURVEYS OF LOCAL GOVERNMENTS

The primary basis for depicting the experiences of local governments in responding to the policies under study involved surveys of local governments in Florida, New Zealand, and New South Wales. These were supplemented with data from respective national decennial censuses (1990 in the United States; 1991 in Australia and New Zealand) and from other secondary sources. The questionnaires that were administered to local governments were comparable, subject to differences in country-specific terminology and detail of questions, in the types of questions that were asked. The following discusses sampling procedures and potential non-response biases.

Sampling procedures

Because the survey data for Florida's local governments were collected as part of an earlier research study, there is a difference in sampling from that employed for the survey data for local governments in New South Wales and New Zealand. The primary difference is that the Florida data were limited to a sample of coastal jurisdictions while the data from the other settings involved

232

complete enumeration of all local governments. Our analyses show that coastal location, when analyzed in New South Wales and New Zealand, does not have a discernible effect on the research findings. The sampling approaches for each survey are explained in the following paragraphs.

Florida sample The survey of local governments in Florida was conducted in 1990. It involved a mail-back survey, with in-person or phone follow-up, that addressed local growth-management planning and other hazard mitigation efforts for a random sample of thirty local governments. The sample was limited to coastal jurisdictions, defined as counties bordering ocean or estuarine (tidal) shorelines and municipalities located within those counties. The sample was drawn from cities and counties with 1990 estimated populations greater than 2,500, excluding Miami (since it would be less comparable with the rest of the jurisdictions in the sample). The sample included twenty-seven municipalities and three counties. Responses were obtained from each of the thirty jurisdictions in the sample. The survey was conducted by staff of the Center for Environmental and Urban Problems under the joint auspices of Florida Atlantic University and Florida International University.

New South Wales sample The survey of local governments in New South Wales, Australia sought a complete enumeration of local councils with risks posed by riverine or surface-water flooding. Mail-back questionnaires were administered to all local councils for which there was a potential flood problem. Personnel from the state agencies responsible for managing the state's flood policy helped develop a list of 155 local councils (out of a total of 177 councils in New South Wales) that fit this criterion. Among these, responses were received from 127 councils for an 82 percent response rate. The survey, with appropriate follow-up, was conducted from April to August 1993 by staff of the Centre for Resource and Environmental Studies of the Australian National University.

New Zealand sample We sought a complete enumeration of local governments in New Zealand through use of two mail-back questionnaires. The first survey addressed local district councils and included four unitary authorities that share functions of both regional and district councils. Responses were received from fifty-nine of the seventy-three local authorities for a response rate of 81 percent. The second survey was administered to twelve regional councils and also entailed collection of regional-level information from the four unitary authorities. Responses were received from each of the regional councils and unitary authorities for a 100 percent response rate. These surveys, with appropriate follow-up, were conducted from April to August 1993 by staff of the Centre for Environmental and Resource Studies at the University of Waikato, New Zealand.

Assessing non-response biases

Table A.1 presents summary information about local governments in New South Wales and New Zealand. The table provides a comparison of data for those jurisdictions from which survey responses were received with corresponding data for those from which responses were not obtained. (Responses were received for all local goverments in the sample in Florida.) Except for ratings of risks posed by selected natural hazards, the comparisons are based on relevant census data. The risk ratings are subjective ratings undertaken by the research team in each country, based on their knowledge of risks posed by various natural hazards.

The local councils in New South Wales that did not respond to the survey have a lower average risk and smaller populations than the corresponding values for those that responded. The differences in mean population growth rates and land area are not statistically significant, but indicate that the non-response consisted mainly of smaller jurisdictions. Overall, this comparison suggests that there is a slight bias in the New South Wales local survey results toward higher-risk, larger jurisdictions. Given their profile, the non-responding councils are likely to have lower commitment and capacity for floodplain management.

The local councils in New Zealand that did not respond also tend to have smaller populations, lower population growth, and smaller physical area than the corresponding values for the local councils that did respond (recognizing that the statistical significance of these differences is weak). Except for flood risks, the non-responding councils differ little on average from the responding councils with respect to risk. The flood risk is on average higher for the non-responding councils, which might lead to a slight understatement in our analysis of the commitment to undertaking hazard-mitigation programs. However, these differences are not great.

CASE STUDIES FOR SELECTED LOCALITIES

In order to illustrate different approaches that local governments have developed for aspects of environmental management and to understand the range of experience under cooperative intergovernmental policies, we undertook a set of case studies in New South Wales and in New Zealand. These are reported in Chapter 8 and Chapter 9. As explained in the following paragraphs, the selection and use of the cases differed somewhat between the two settings.

Case studies in New South Wales Five local councils were selected in order to provide a contrast between "leading" and "lagging" jurisdictions with respect to hazards management and compliance with the state's flood policy. Several criteria were used to select the cases. Small, rural councils with populations of a few thousand and for which responses to the local government survey indicated minimal flood exposure were excluded. The remaining local governments were

Table A.1 Non-response analysis for surveys of local governments

Comparison item	Mean values for local governments:[a]		
	Responding to the survey	Not responding to the survey	P-value[b]
New South Wales Local Councils			
Number of cases	127	28	–
Flood risk index (coded 1= no threat, 2= small threat, 3= moderate threat, 4= severe threat, 5= very severe threat)	3.07 (1.10)	1.82 (.82)	<.01
1991 Population	37,780 (49,851)	16,512 (21,871)	.03
Population growth 1981 to 1991 (percentage change)	14.75 (25.18)	6.17 (25.44)	.13
Land area (sq. km)	4,491 (8,012)	3,571 (4,320)	.56
New Zealand district councils			
Number of cases	59	14	–
Flood risk (coded 1= low or none, 2= moderate, 3= high risk)	2.53 (.86)	3.00 (.00)	.04
Earthquake risk (coded 1= low or none, 2= moderate, 3= high risk)	2.36 (.94)	1.93 (1.00)	.14
Coastal hazard risk (coded 1= low or none, 2= moderate, 3= high risk)	1.58 (.91)	1.90 (.88)	.31
1991 Population	49,551 (62,505)	31,670 (26,551)	.30
Population growth 1981 to 1991 (percentage change)	7.53 (13.68)	.41 (17.28)	.10
Land area (hectares)	398,357 (519,554)	290,899 (219,988)	.45

Sources: Compiled by authors from secondary data
Notes:
   [a] Cell entries are mean values with standard deviations in parentheses.
   [b] t-test of difference of means between results for responses and non-responses.

ranked in terms of the combined measures from the local government surveys of efforts to manage floods, commitment of local government officials to addressing flood problems, and reported amount of development at risk of flooding.

Cases were selected from the upper and lower quartiles of these rankings on a judgmental basis, while also taking into account comparability in growth rates and population size. An additional criterion that was particularly relevant to our discussion of changing risk profiles in Chapter 9 was the availability of data about the nature of development within a community prior to the introduction of the state's merits flood policy in 1985. These selection criteria resulted in two councils that are considered "leaders" (Fairfield and Muswellbrook), two with intermediate status (Lismore and Singleton), and one "laggard" (Liverpool). Descriptive information about each is provided in Chapter 9.

The case studies in New South Wales involved interviews within each locality with participants in the floodplain-management process, typically involving the public works engineer, a local planner, elected officials, and representatives of community groups. These interviews provided an understanding of the local context. The more labor-intensive aspects of the case studies involved characterization of changing development patterns based on an inventory of flood-prone residential structures undertaken in 1993 and 1994. In most cases, we were able to compare this inventory with parallel inventories undertaken prior to the implementation of the state's merits flood policy.

The construction of databases about buildings at risk from flooding was based on the methodology developed for use with a computerized damage-assessment program, ANUFLOOD, developed at the Centre for Resource and Environmental Studies of the Australian National University. (References to this program are cited in Chapter 9.) This program is widely used by government agencies in New South Wales, and elsewhere in Australia, to assess urban floodplain damages and possible future mitigation measures. The comparison of development profiles before and after the introduction of the state's flood policy in 1985 allows precise discussion of the outcomes for both new development and for removal and reduction of risk that have taken place within flood-prone localities. Such data, however, only provide a partial answer to the counterfactual of what might have happened if the policies had not been in place.

Case studies in New Zealand Two district councils were selected in New Zealand in order to provide a contrast between "leading" and "lagging" approaches to hazards management and a contrast in responses to the *Resource Management Act*. Similar procedures were employed as with the New South Wales data in using the local survey responses to identify localities that could be considered leaders in responding to the new policy mandate. With these results in mind and considering access to historical data about local policy change over time in managing flood hazards, cases were selected on a judgmental basis. This led to the selection of Thames as an example of a local government undertaking innovative responses and Invercargill as an example of a location for which the response was more muted. The case studies involved interviews with staff of the district councils and with staff from the corresponding regional council. Historical data about flood experience and risks were collected, but there was insufficient basis to develop the detailed profiles of development patterns that were developed for selected localities in New South Wales.

STATE AND CENTRAL GOVERNMENT AGENCIES

In order to characterize key aspects of the intergovernmental implementation of the policies under study, we undertook surveys of relevant state and central government agencies, and conducted interviews with key personnel in each agency. The surveys entailed mail-back questionnaires that addressed funding, staffing, level of effort for various implementation activities, and respondent assessments of various dimensions of state (or central government) policy implementation. The interviews consisted of in-depth probing about policy evolution and implementation. The state and central government data were supplemented with surveys of regional offices of state agencies in New South Wales and in New Zealand. In each of these settings, there were several rounds of interviews over a two-year period.

The Florida data collection and interviews involved key personnel from the state's Department of Community Affairs, the lead agency for administering the state's growth management and environmental legislation, in conjunction with the earlier research study in 1990. The state-level data collection in New South Wales involved the two lead agencies responsible for implementing the state's merits flood policy – the Department of Public Works and the Water Resources Department. Similar information was collected in New Zealand from the national Ministry for the Environment and the Department of Conservation. The data for New South Wales and New Zealand were collected in 1994.

As discussed in Chapter 5, we sought to expand our understanding of intergovernmental influence by examining the influence of regional offices of relevant agencies. In New South Wales, we collected data from five of the six regional offices of the Department of Public Works and from eight of the nine regional offices of the Water Resource Department. (Both of the non-responding offices are in areas with relatively low flood risks.) We attempted to collect corresponding data for regional offices of New Zealand's agencies, but encountered gaps in responses. We received partial responses from three of the four regional offices of the Ministry for the Environment and for twelve of the fourteen conservancies of the Department of Conservation. The information for regional offices is based on questionnaires filled out by lead personnel, as supplemented in some cases by in-person interviews. The questionnaires addressed regional office capabilities, commitment, and approach to dealing with local governments as well as factual information about staffing and the extent of contact with local governments within their agency purview.

APPENDIX

MEASURES AND DATA ANALYSES

The measures that we employ as part of our statistical analyses are primarily derived from the local government and state (or central government) agency surveys. Many of the measures we employ are indices constructed from multiple items in the questionnaires. Where comparisons between settings were involved, we attempted to obtain as much comparability in index construction as possible while taking into account differences in question emphasis and item phrasing.

State and central government agency variables Table A.2 summarizes the key variables and their measurement for state agencies in Florida and New South Wales, central government agencies in New Zealand, and relevant regional offices of government agencies. These are relevant primarily to the discussion in Chapter 5 of the translation of policies into practice.

Measures of the quality of regional policy statements Part of the analysis of regional government roles, discussed in Chapter 6, consisted of rating the quality of draft policy statements produced by regional councils and unitary authorities in New Zealand. We obtained all of the policy statements that were completed in draft form (for submission for public review), which involved policies prepared by each of the twelve regional councils and two of the four unitary authorities. The evaluation consisted of a rating of quality with respect to their "goals" – the extent to which goals for hazard management were specified, and to their "policies" – the extent to which specific policies were specified for hazards management. The coding procedure was modified from one originally developed for analyzing local governments' plans in the United States by Philip Berke of the University of North Carolina at Chapel Hill, who assisted in the design of this portion of the research. Using a common coding form, each policy statement was rated for the presence or absence of a series of potential hazard-management goals and policies.

The goals dimension involved six potential hazard-related goals (reducing property loss, protecting safety of the population, reducing damage to public property, minimizing fiscal impacts of disasters, distributing hazard management costs equitably, and promoting hazard awareness) and three potential environment-related goals (reducing hazards to achieve preservation of natural areas, reducing hazards to preserve open space and recreation areas, and reducing hazards to achieve good water quality). Each policy statement was evaluated for the presence or absence of these items and scored one for each item mentioned and zero if not mentioned.

The policies dimension involved elements of six types of policies for managing hazards: (1) awareness – educational awareness, voluntary real estate hazard disclosure, disaster warning systems, posting of signs to indicate hazardous areas, programs to encourage purchase of flood or earthquake insurance, technical assistance; (2) regulatory policies – permitted land uses, density controls over

238

Table A.2 Variables for state and central government agencies

Variables	Measurement
Agency capacity	Mean of respondent rating (1= poor to 7= excellent for each item) of adequacy of staffing, adequacy of expertise, and access to local government officials. Higher scores reflect greater perceived capacity.
Agency commitment	Mean of respondent rating (1= low to 7= high for each item) of agency endorsement of policy, importance of policy relative to other agency activities, willingness of leadership to promote policy, and seniority/influence of individuals implementing the policy. Higher scores reflect greater perceived commitment.
Agency implementation effort	Measure of implementation effort applied to regional offices, calculated from regional office surveys as staff effort (full-time-equivalent staff times percentage of time spent on implementation tasks) divided by number of local governments within the purview of the regional office.
Agency implementation style	Index based on mean respondent rating (1 to 7 scale for each item) of agency communications with local governments (written to less formal), interpretation of policy guidelines (as strict rules to flexible interpretation), and substantive emphasis (strict adherence to policy goals). Scaled so that low scores indicate a formal, legalistic style and high scores indicate an informal, facilitative style.
Hours of assistance	Mean number of hours of technical assistance (defined below) provided by agency staff in the most recent year to each local government with its purview.
Number of jurisdictions	Number of local governments within the purview of the agency.
Staffing ratio	Ratio of the number of staff in the agency to the number of local governments within the purview of the agency.
Technical assistance index	Index computed as the amount of technical assistance provided to each jurisdiction within the purview of an agency. Calculated as the sum of a set of possible agency actions (each 1= yes, 0= no; possible range= 0 to 6): responded to questions, reviewed plan or policy, distributed guidance materials, distributed example plan or policy, telephone consultation, and on-site technical assistance.

land use, transfer of development rights, cluster development provisions, set-back requirements, site review requirements, special study or impact assessment requirements, building standards for hazardous locations, mandatory real estate hazard disclosure, land and property acquisition through eminent domain, financing for mitigation, and mandatory retrofitting of structures for hazard resilience; (3) incentive policies – voluntary retrofitting of private structures, voluntary land and property acquisition, tax abatement for employing mitigation measures, density bonus, low interest loans for retrofitting buildings; (4) infrastructure policies – structural controls over hazards, capital improvement adjustments, retrofitting public infrastructure, and provisions for critical facilities; (5) recovery planning provisions – post-disaster land-use changes, post-disaster building design changes, moratorium on reconstruction, establishment of a recovery organization, capital improvement adjustments, provisions for acquisition and relocation of damaged properties, identification of mechanisms to finance recovery provisions; and (6) emergency planning provisions – evacuation provisions, sheltering provisions, emergency plan preparation. Each policy statement was scored with a score of zero if the policy provision was not present, one if it was a recommended provision, and two if it was a mandatory provision.

Scores were computed for each policy statement along each dimension with appropriate standardizing to set the minimum score of zero and the maximum score of 100. This involved summing raw scores for the goal and policies dimensions for each of four possible types of hazards – flooding, coastal storms, earthquakes, or landslides. These were then divided by the number of hazards present in the community (in order to adjust for differences in hazard vulnerability), and further divided by the number of potential items in each dimension (in order to provide a common metric). For ease of presentation, final scores were multiplied by 100. Inter-coder reliability estimates for a sub-sample involving four regional policy statements showed agreement among two coders on 92 percent of all items.

Local government variables Table A.3 summarizes the key variables and their measurement. Given that much more detail was collected for the surveys of local governments in New South Wales and New Zealand than for the survey of local governments in Florida, more of the variables apply to the former settings than the latter. The local government variables are relevant to analyses presented in Chapters 6 through 10 of the book.

Data analyses and caveats

The analyses consist mainly of comparisons of the impacts of different forms of intergovernmental policies on local approaches to aspects of environmental management. We employ descriptive statistics where appropriate and report in endnotes the results of regression and other multivariate analyses that were

240

Table A.3 Variables for local governments

Variables	Measurement
Amount of hazard-free land	Index based on the respondent rating of the amount of undeveloped land available in areas that are not hazard-prone (scaled 1 = none to 5 = extensive amount). Based on assessments of flood hazards in New South Wales and Florida, and assessments of flood, coastal, and earthquake (ground-shaking) hazards in New Zealand. Higher scores indicate relatively larger areas of hazard-free land.
Capacity of the local government	Index of overall capacity of the local government to carry out hazard (or environmental) management programs. Based on mean of respondent ratings of the adequacy of funds for addressing the problem, access to in-house technical expertise, and ability to enforce relevant rules or regulations (scaled 1 = poor to 7 = excellent). Higher scores indicate greater perceived capacity.
Catastrophic losses from disasters	Respondent rating of the impact of a disaster occurring within the past twenty years. Catastrophic events are considered as those rated as having an "enormous impact – devastation was widespread and severe" or a "large impact – devastation was widespread and moderate."
Change in commitment of elected officials and staff	Respondent rating of the change in commitment of elected officials, and separately of staff to hazard-management programs after introduction of a cooperative mandate. Rating was a 3-point scale of an increase, no change, or decrease.
Change in exposure to natural hazards	Respondent rating of the change in the exposure of people or property to risks posed by natural hazards (or flood hazards) after introduction of a cooperative mandate. The rating was a 3-point scale of an increase, no change, or decrease.
Change in fairness of decision-making about development	Respondent rating of the change in the fairness of decision-making by the local government concerning development in hazard-prone areas after introduction of a cooperative mandate. The rating was a 3-point scale of an increase, no change, or decrease.
Change in flexibility of rules governing development	Respondent rating of the change in flexibility of rules or regulations concerning development in hazard-prone areas after introduction of a cooperative mandate. The rating was a 3-point scale of an increase, no change, or decrease.
Change in willingness to negotiate development	Respondent rating of the attitude change of the staff of the local government in their willingness to negotiate with developers or others who want to build in hazard-prone areas. The rating was a 3-point scale of an increase, no change, or decrease.

241

Table A.3 Variables for local governments (cont'd)

Variables	Measurement
Character of building regulations	Rating of the "strictness" or "flexibility" of building regulations that have been adopted by a local government, based on respondent identification of types of regulations the jurisdiction has adopted. Strict building regulations are rules that prescribe particular practices. Flexible building regulations are rules that specify performance standards allowing for different means of adherence, or rules that specify criteria for making decisions about acceptable buildings. Dichotomous coding as either "strict" or "flexible."
Character of land-use controls	Rating of the "strictness" or "flexibility" of land-use controls that have been adopted by a local government, based on respondent identification of types of rules the jurisdiction has adopted. Strict land-use controls are rules that prohibit or otherwise restrict development in hazard-prone areas. Flexible land-use controls are rules that permit low levels of development, or tradeoffs in return for development rights in hazard-prone areas. Dichotomous coding as either "strict" or "flexible."
Commitment of elected officials	Respondent rating of the commitment of elected officials to hazard (or environmental) management programs (scaled 1 = poor to 7 = excellent). Higher scores indicate greater perceived commitment.
Commitment of the local government	Index of overall commitment of the local government (or environmental) management programs. Based on mean of respondent ratings of the commitment of elected officials and staff (scaled 1 = poor to 7 = excellent). Higher scores indicate greater perceived commitment.
Commitment of staff	Respondent rating of the commitment of senior staff of the local government to hazard (or environmental) management programs. Based on mean of respondent ratings of staff commitment and staff influence (scaled 1 = poor to 7 = excellent). Higher scores indicate greater perceived commitment.
Constituency demands	Index measuring extent of different constituency activities within the local government addressing hazard problems. The relevant groups are: local business, environmental, neighborhood, and individuals not associated with any particular group. For each group an index was created based on whether or not each of four possible actions had been taken (1 = yes, 0 = no): asked for information; became active in asking for community action; attended meetings; and/or served on relevant committees addressing the problem. The overall index of constituency demands is the sum of the scores for the

Table A.3 Variables for local governments (cont'd)

Variables	Measurement
Constituency demands (cont'd)	four groups for a possible total score of sixteen. Higher scores indicate greater constituency demand to address hazard-related problems.
Demand for development in hazardous areas	Index based on the respondent rating of the demand for development in hazard-prone areas (scaled 1 = none to 5 = extensive). Based on assessments of flood hazards in New South Wales and Florida, and assessments of flood, coastal, and earthquake (ground-shaking) hazards in New Zealand. Higher scores indicate greater demand for such development.
Effort to build awareness	Respondent rating of the degree of effort the local government is putting into increasing people's awareness of hazards (scale 1 = low to 7 = high). Higher scores indicate greater efforts. Data are reported for Florida and New South Wales only. These were recoded for some analyses into fewer categories.
Effort to manage development	Respondent rating of the degree of effort the local government is putting into land use measures to limit development in hazard-prone areas or into requirements to make development in hazard-prone areas safer (scale 1 = low to 7 = high). Calculated as the mean of respondent rating of each item, for which higher scores indicate greater efforts.
Floodplain management committee formed	Respondent reporting of whether or not the recommended floodplain management committee was formed (New South Wales only; coded 1 = yes, 0 = no).
Government expenditures per capita	1991/92 local government expenditures per capita; based on secondary sources.
Influence of developers	Respondent rating of the influence of developers on the decisions about development by elected officials of that local government (1 = not influential to 7 = highly influential). Higher scores indicate greater perceived influence.
Influence of neighborhood groups	Respondent rating of the influence of neighborhood groups on the decisions about development by elected officials of that local government (1 = not influential to 7 = highly influential). Higher scores indicate greater perceived influence.
Intergovernmental advice or assistance	Respondent ratings of the extent to which local governments relied on advice from regional councils, other local governments, and relevant agencies (each scaled 1 = rarely to 7 = often). Respondents also rated the relevance of the advice from each source (scaled 1 = rarely use to 7 = often use). Higher scores indicate greater reliance and greater perceived relevance.

Table A.3 Variables for local governments (cont'd)

Variables	Measurement
Liability considerations for hazards management	Respondent rating of the effect that the potential for legal liability had on the local government's hazard-management program (0= no effect to 3= played a major role). Higher scores indicate greater influence.
Perception of state (or national) agency implementation style	Respondent rating of different aspects of the style of the relevant state (or national) agencies in dealing with the local government. The style was rated with respect to several items (each on a 1 to 7 scale, endpoints anchored by labels in parentheses): substantive emphasis (rules vs. goals), cooperation (limited vs. extensive), and persuasion (sanctions vs. incentives). Scaled so that low scores indicate perceptions of a formal, legalistic style and high scores indicate perceptions of an informal, facilitative style.
Perceptions of decision-making outcomes	Respondent rating of whether the newly mandated planning process resulted in an increase, no change, or decrease in each of the following: the flexibility in rules or regulations concerning development; local government willingness to negotiate with developers; fairness of decision-making about development in hazardous areas. Respondents were also asked how important the relevant policy was in influencing the reported change (1= no importance to 4= very important).
Plan quality	Index of the quality of regional policy statements in New Zealand.
Population	Population data from 1991 censuses (New South Wales and New Zealand), 1990 census (Florida).
Population growth rate	Percentage growth in population from 1981 to 1991 (New South Wales and New Zealand), 1980 to 1990 (Florida).
Potential losses from disaster	Recoding of respondent rating of demand for development in hazard-prone areas into three categories (low, moderate, high).
Prior losses from disasters	Respondent reporting of whether or not local governments reported more than incidental damage from a given set of events in the past twenty years. Based on damaging events from flood hazards in New South Wales and Florida, and damaging events from flood, coastal, or earthquake (ground-shaking) hazards in New Zealand. Coded as a dichotomous variable (1= prior losses, 0= no noteworthy losses).
Procedural compliance with local planning provisions	Whether or not the local government complied with procedural requirements in preparing mandated plans

Table A.3 Variables for local governments (cont'd)

Variables	Measurement
Procedural compliance with local planning provisions (cont'd)	(within specified deadlines, if they existed). Based on survey data in New South Wales, and secondary data for Florida and New Zealand.
Process considerations in local planning	Respondent rating of whether the cooperative policy resulted in an increase, decrease, or no change for each the following: consideration of environmental issues in planning; public participation in the planning process; consideration of extreme flooding in planning; and use of benefit–cost analysis in planning assessments. Respondents were also asked to rate how important the newly mandated planning process was in influencing the reported change (1 = no importance to 4 = very important).
Risk of earthquake	Respondent rating of the threat posed to people and property within that local government jurisdiction by earthquake hazards (1= none to 5= very severe). The higher the score, the greater the perceived risk posed by earthquakes.
Risk of flooding	Respondent rating of the threat posed to people and property within that local government jurisdiction by flooding by rivers (1= none to 5= very severe). Asked separately in New South Wales for flooding within the 100-year flood level, above the 100-year flood level, and by stormwater surface flooding. The higher the score, the greater the perceived risk posed by floods.
Staff professionalism	Measure based on the percentage of planning staff who are certified by relevant professional planning organizations. Higher scores indicate greater professionalism.
Stage in merits planning process	Respondent rating of the stage that the local government has reached in the prescribed planning process for floodplain management in New South Wales. Recoded so that 1 = not started, or only formed committee; 2 = flood study or flood standard completed; 3 = management or issue study completed; 4 = management plan or implementation plan completed.
Strategy for hazard management	Respondent identification of actions undertaken by the local government to address relevant hazards. Building regulations defined as controls on types of building or other requirements for building in hazard-prone areas. Land-use regulations defined as restrictions or other rules governing development patterns in hazard-prone areas. Structural-protection measures defined as use of engineered works to reduce hazard potential.

undertaken to control for the effects of situational variables. Where analyses involve data from more than one setting in calculating a given statistic, cases are weighted to give equal weight to each case (local government) so that differences in sample sizes do not bias aggregate statistics. For most analyses, however, data are reported separately for Florida, New South Wales, and New Zealand.

There are extraneous factors that potentially are not controlled by the research design and statistical measurements we employ. In particular, differences between Florida's coercive policy impact and the cooperative intergovernmental policy impact in New South Wales and New Zealand can potentially be explained by factors we do not measure or adequately control. Factors with this potential include the substantial exposure of local governments in Florida, but not New South Wales or New Zealand, to hurricane hazards and the much greater growth pressures experienced in Florida, where average rates of population growth over the 1980s were more than double those in New South Wales or New Zealand. In combination, those two factors could create a climate in Florida much more conducive to local government growth management than existed in New South Wales or New Zealand. The effects we attribute to the Florida planning mandate could in fact be a product of the policy environment rather than the policy studied. However, we think our controls for risk exposure and growth rates adequately address these threats to internal validity.

Another caveat stems from the fact that most of the data are cross-sectional. They provide an understanding of local governments' effort, commitment, and capacity after policies have been in place for up to eight years. What we attribute as policy influences of the policies may be the legacy of prior policies and our finding of differential policy effects may stem in part from different starting points. (These, of course, are standard problems of the use of *ex post facto* controls.) We partially address these concerns through longitudinal analyses based on retrospective data we collected in New South Wales and in New Zealand, and through use of case studies to understand changing development patterns over time in selected settings. We have been very self-conscious of data limitations and note them where relevant in the book.

INDEX

Page numbers in *italics* denote illustrations.

123, 136; *see also* environmental assessment processes
Monroe County (Florida) 27
Muswellbrook (New South Wales) 163, 166, *167*, 168–9, 175, 181–3, 236

National Development Act 1979 (New Zealand) 50
National Flood Insurance Program (USA) 28, 32, 148, 149, 152, 206
National Landcare Programme, Australia *see* Landcare Programme
national park system, New Zealand 65
National Water and Soil Conservation Authority (New Zealand) 152, 158
natural hazards *see* hazards
New South Wales, land use *71*; *see also* floodplain management policy, New South Wales
New South Wales Flood Mitigation Grants Act 73
New Zealand: erosion *113*; flooding *148*; government reform 14, 43, 47–51, 63–4, 65; land use *44*; *see also* environmental management policy, New Zealand
New Zealand Business Roundtable 103–4, 135
New Zealand Forest Service 48
New Zealand Planning Institute Conference 135

Omnibus Growth Management Act (Florida) 31, 32
Otago Regional Council (New Zealand) 122
outcomes *see also* development: changes in decision-making; exposure

Palmerston North (New Zealand), floodplain management *54*
Parliamentary Commissioner for the Environment (New Zealand) 48–9, 55
Pelham, Thomas 27
penalties 3, 46, 47, 55; *see also* sanctions
Petak, William 196
planning: Florida 8, 26–7, 28–9, 30–40; local 126–44, 217, 226, 245; New South Wales 10–11, 69–70, 72–85; New Zealand 9, 43, 45–64; regional 110, 111, 112, 114, 116–21, 123; requirements compared 92–3; *see also*

development, management of; land-use control; local plans
Planning and Environment Commission (New South Wales) 73
Planning Tribunal (New Zealand): decisions from 56, 114, 121, 141, 157–8; role of 52, 55, 59, 129, 135, 226
policy change 91, 221; Florida growth management program 36–7; New South Wales flood policy 76–7; New Zealand environmental management 62–3
policy design: intergovernmental policies 3–5, 90–4, 215; *see also* floodplain management policy, New South Wales; growth management program, Florida; *Resource Management Act 1991*, New Zealand
policy learning 215–16
pollution management 49, 51, 52, 57, 108
Powell, David 139
procedural compliance 4, 130–6, 143–4, 244–5
property rights 1, 13, 38, 75, 147
Prospect Creek 186–8
public information campaigns 9, 204–7, 210, 227
public participation 127, 128, 207, 221–2, 226–7, 242–3, 245; Florida 34; New South Wales 79, 81, 93, 138, 168, 187; New Zealand 52, 56, 59, 64, 131, 138, 162; *see also* interest groups

Queensland 71

regional government: data collection, methodological notes 233, 237–9, 243; Florida 27, 30–1, 34, 37, 39–40, 108, 111; New South Wales 79, 108, 224; New Zealand 9, 14, 45–50, 52, 54–9, 61, 64–5, 103, 108–24; role of 92–3, 108–24, 136, 215–16, 222–3, 224–5; *see also* Canterbury Regional Council; Otago Regional Council; Southland Regional Council; Waikato Regional Council; Wellington Regional Council
regional planning councils *see* regional government